Observing the Universe: A Guide to Observational Astronomy and Planetary Science

Compiled by a team of experts from The Open University, this textbook introduces a range of techniques and skills designed for students who are undertaking observational work in astronomy and planetary science.

Part I introduces modern tools and techniques, starting with a discussion of the night sky and an introduction to positional astronomy. This is followed by a description of telescopes, spectrographs and astronomical detectors. Several chapters are devoted to a discussion of CCD data-reduction techniques, and how to deal with photometric and spectroscopic data. The use of petrographic microscopes for the analysis of extraterrestrial mineral samples is introduced, and some guidance on the interpretation of images of planetary surfaces is given.

Part II focuses on the development of skills that are of relevance to astronomy and planetary science projects. There are chapters on teamwork, preparing for practical work and keeping records. This is followed by a review of how to deal with experimental uncertainties, notes on analysing experimental data, and guidelines on making use of graphs and computers. The book concludes with a discussion of how best to communicate the results of scientific investigations.

Written in an accessible style that avoids complex mathematics, and illustrated in colour throughout, this book is suitable for self-study. It contains numerous helpful learning features such as chapter summaries, student exercises with full solutions, and a glossary of terms.

Observing the Universe is directed at students who are about to engage in observational optical astronomy and planetary science, but who have had little exposure to this discipline before. The book will also appeal to amateur astronomers, to beginners as well as to experienced backyard observers, who would like to go beyond the passion of pure observing into the realm of scientific measurements.

About the editor:

ANDREW NORTON studied for a Ph.D. at the University of Leicester in the area of observational X-ray astronomy. During this time he worked on data from the *EXOSAT* and *Ginga* satellites, concentrating on interacting binary stars containing white dwarfs or neutron stars. He then spent four years at the University of Southampton where his research became multi-wavelength, extending to include optical, infrared and radio observations. Whilst there, he was involved with running an undergraduate observational astronomy course based in Tenerife. Dr Norton joined The Open University in 1992, where he is now a Senior Lecturer in Physics and Astronomy. His research interests continue in the area of multi-wavelength observations of interacting binary stars and time-domain astrophysics. He has authored teaching materials in Physics and Astronomy across the undergraduate curriculum, including ten OU study texts, as well as video programmes, CD/DVD-ROM packages and experimental projects for OU residential schools.

Background image: Star trails in the southern sky. A time-exposure of the night sky obtained at Siding Spring Observatory, New South Wales. The dome in the foreground is that of the Anglo-Australian Telescope. (Courtesy of the Anglo-Australian Observatory/David Malin Images)

Thumbnail images: (from left to right) The 400 optical fibres on the 2dF instrument at the Anglo-Australian Telescope (Courtesy of the Anglo-Australian Observatory); an example of a mineral viewed through polarizing filters in a petrographic microscope; a 40 cm diameter Schmidt-Cassegrain telescope, of the type often used for student projects; a CCD frame of a star field, with a typical stellar spectrum superimposed.

Observing the Universe:
A guide to Observational Astronomy and Planetary Science

Edited by Andrew J. Norton

Authors:

W. Alan Cooper
Ian A. Franchi
Stuart M. Freake
Simon F. Green
Carole A. Haswell
Barrie W. Jones
Ulrich C. Kolb
T. J. Lowry McComb
Andrew J. Norton
David A. Rothery
Sean G. Ryan

The Open University

CAMBRIDGE
UNIVERSITY PRESS

PUBLISHED BY THE PRESS SYNDICATE OF THE UNIVERSITY OF CAMBRIDGE

The Pitt Building, Trumpington Street, Cambridge, United Kingdom

CAMBRIDGE UNIVERSITY PRESS

The Edinburgh Building, Cambridge, CB2 2RU, UK

40 West 20th Street, New York, NY 10011–4211, USA

477 Williamstown Road, Port Melbourne, VIC 3207, Australia

Ruiz de Alarcón 13, 28014 Madrid, Spain

Dock House, The Waterfront, Cape Town 8001, South Africa

http://www.cambridge.org

First published 2004

Copyright © 2004 The Open University

All rights reserved. No part of this publication may be reproduced, stored in a retrieval system, transmitted or utilized in any form or by any means, electronic, mechanical, photocopying, recording or otherwise, without written permission from the publisher or a licence from the Copyright Licensing Agency Ltd. Details of such licences (for reprographic reproduction) may be obtained from the Copyright Licensing Agency Ltd of 90 Tottenham Court Road, London W1T 4LP.

Open University course materials may also be made available in electronic formats for use by students of the University. All rights, including copyright and related rights and database rights, in electronic course materials and their contents are owned by or licensed to The Open University, or otherwise used by The Open University as permitted by applicable law.

In using electronic course materials and their contents you agree that your use will be solely for the purposes of following an Open University course of study or otherwise as licensed by The Open University or its assigns.

Except as permitted above you undertake not to copy, store in any medium (including electronic storage or use in a website), distribute, transmit or re-transmit, broadcast, modify or show in public such electronic materials in whole or in part without the prior written consent of The Open University or in accordance with the Copyright, Designs and Patents Act 1988.

Edited, designed and typeset by The Open University.

Printed and bound in the United Kingdom by Bath Press, Blantyre Industrial Estate, Glasgow G72 0ND, UK

A catalogue record for this book is available from the British Library

ISBN 0 521 60393 5

This publication forms part of an Open University course SXR208 *Observing the Universe*. Details of this and other Open University courses can be obtained from the Course Information and Advice Centre, PO Box 724, The Open University, Milton Keynes MK7 6ZS, United Kingdom: tel. +44 (0)1908 653231, e-mail general-enquiries@open.ac.uk

Alternatively, you may visit the Open University website at http://www.open.ac.uk where you can learn more about the wide range of courses and packs offered at all levels by The Open University.

To purchase a selection of Open University course materials visit the webshop at www.ouw.co.uk, or contact Open University Worldwide, Michael Young Building, Walton Hall, Milton Keynes MK7 6AA, United Kingdom for a brochure. tel. +44 (0)1908 858785; fax +44 (0)1908 858787; e-mail ouwenq@open.ac.uk

1.1

CONTENTS

INTRODUCTION

Astronomy and planetary science are observational sciences that involve investigating the workings of the Universe. They encompass cosmological observations on the scale of the entire Universe, observations of the planets in our local Solar System, and everything in the intervening range.

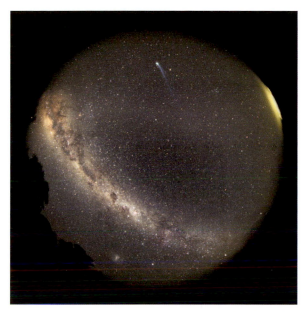

Figure 0.1 A view of the southern sky dominated by the band of the Milky Way. Comet Hyakutake is also visible. (© *Gordon Garrard*)

From a dark site, the view of the night sky can be truly awe-inspiring (Figure 0.1). For many people, merely observing the splendours of the Universe in this way is enough. Others may wish to see more, through telescopes and binoculars, and a few people will wish to *understand* a little about the objects they can see. This guide is intended for those who wish to go beyond mere stargazing and begin to make scientific observations of planets, stars and galaxies. In particular, it is designed to provide university students in astronomy and planetary science with the requisite knowledge needed prior to a session at an astronomical observatory or planetary science laboratory.

Since the time of Galileo (Figure 0.2), astronomers have designed telescopes and used them to investigate astronomical bodies, from the Moon to quasars at the limits of the observable Universe. Furthermore, astronomical observations now use the whole range of the electromagnetic spectrum, from radio waves through to γ-rays (pronounced 'gamma rays'); telescopes are operated both from the ground and from satellites in orbit around the Earth (Figure 0.3a–0.3d); space probes carrying a variety of detectors visit bodies throughout the Solar System (Figure 0.3e); and samples returned from the Moon and meteorites can be analysed in the laboratory. The wealth of data that astronomers and planetary scientists collect provide evidence to answer questions about topics including how the Solar System formed, whether planetary systems exist around other stars, the life history of stars and galaxies, and how the Universe originated in the Big Bang.

Figure 0.2 Galileo Galilei (1564–1642) was the first person to use a telescope to observe the Universe. He discovered the four largest moons of Jupiter, observed sunspots, and discerned individual stars comprising the Milky Way.

(a)

(b)

(c)

(d)

(e)

Figure 0.3 Telescopes and astronomical satellite observatories operate across the electromagnetic spectrum, and space probes carry a variety of detectors to distant parts of the Solar System. (a) The Hubble Space Telescope is a satellite-based observatory operating in the infrared/optical/ultraviolet. (© *NASA*) (b) The James Clerk Maxwell Telescope is a mm-wave telescope in Hawaii. (*Image courtesy of the James Clerk Maxwell Telescope, Mauna Kea Observatory, Hawaii*) (c) The Very Large Array is an array of radio telescopes in New Mexico. (© *NRAO/AUI/NSF*) (d) The XMM-Newton satellite is an X-ray observatory operated by the European Space Agency (ESA) (© *ESA*). (e) The Cassini–Huygens spacecraft is a mission to Saturn and its largest satellite Titan, carrying a range of scientific experiments. (© *NASA*)

So astronomical observations have played – and continue to play – a crucial role in developing our understanding of the Universe. If you want to understand and get a feel for the role of astronomical observations, perhaps the best way is to perform some yourself.

Throughout this book, we use the term **observatory** to denote a ground-based facility that has equipment for carrying out astronomical observations in the *optical* waveband and equipment for the analysis of extraterrestrial samples. Likewise, for convenience, the term **astronomer** is used to denote any scientist researching in astrophysics, cosmology or planetary science.

Broadly speaking, each project that you carry out at an observatory will have several phases. The first phase is *planning* the observations or investigations that you will carry out; the second phase will often involve *acquiring the data* using a telescope; the third phase will usually involve working on a computer to *reduce and analyse the data* you have obtained; the fourth phase will be the *interpretation* of your results where you draw conclusions; and the final phase will generally involve *writing-up* your project and/or *presenting* the results to others.

These stages parallel those undertaken by astronomers engaged in research. As a student, by following similar processes, you will be learning to think like an astronomer, and this will give you insights into tackling practical problems that will be valuable in many other areas of work experience.

Various other benefits come from engaging in observational astronomy and planetary science as a student. Many projects demonstrate important concepts and phenomena of physics, and seeing a phenomenon or measuring an effect can make it more real and memorable. Carrying out astronomical observations develops a wide range of skills that you will use in other areas besides astronomy and planetary science – everything from planning to problem solving, from analysing data to presenting results. Observational astronomy work also gives an opportunity to work collaboratively with other students, and this can be a powerful learning experience, as well as developing social skills.

In this book, we will discuss a variety of aspects of observational astronomy and planetary science, with the aim of providing a sound basis for tackling projects in practice. The book is divided into two parts.

Part I: *Techniques* begins by introducing you to concepts in positional astronomy and discusses how to locate the objects that you're interested in observing. Chapters 2–4 briefly describe astronomical telescopes, spectrographs and detectors and then Chapter 5 describes the general procedures for processing images obtained with a CCD detector before the data can be analysed. Chapters 6 and 7 discuss the general principles behind astronomical photometry and spectroscopy which will underlie many of the observations that you will carry out. The final two chapters of Part I introduce microscopes and microscopy techniques as vital tools of planetary scientists when studying samples returned from the Moon or meteorites and briefly consider the techniques of interpreting images of planetary surfaces.

Part II: *Skills* concentrates on key skills associated with practical science, discussed in the context of work at an astronomical observatory. It begins, in Chapter 10, by considering aspects of teamwork that are relevant to observational projects in astronomy and planetary science. In Chapter 11 we then discuss the preparation and planning for an observational project. A little time invested in this stage can pay dividends in the quality of your results and the efficiency with which they are obtained. Chapter 12 is about keeping records of your observations – recording what you do, how you do it, what you observe and measure, the analysis and interpretation of your results, and so on. Without any records to refer back to, you are likely to find that the details of how you carried out the observations and any conclusions that you draw from them will be rather ephemeral and soon forgotten. We provide some guidelines for maintaining an observatory notebook, and you may find it helpful to refer to these guidelines when carrying out your observations.

Chapter 13 tackles an important topic for any experimenter who makes measurements: how reliable are the results? Scientists make the reliability of a numerical result explicit by quoting an uncertainty with a measured value; for example, an astronomical magnitude might be quoted as $m_v = 12.3 \pm 0.2$. In this section, the quantification and combination of uncertainties are discussed in detail. Most astronomical and planetary science projects involve numerical analysis of the data, and Chapter 14 gives a few brief guidelines that should help with this. Data analysis may involve graphs, and graphs are frequently used by astronomers as a powerful tool for displaying and interpreting results; Chapter 15 deals with this topic in some detail. Both data analysis and graph plotting have been revolutionized in the last twenty years by the ready availability of graphic calculators and computers. Chapter 16 indicates a few of the ways that these are used in astronomical work, including a brief introduction to spreadsheets.

For practising astronomers and planetary scientists, carrying out observations and analysing the results are generally of limited value unless those results are then communicated to other scientists. This communication is generally achieved by publishing the results in a scientific journal or reporting them at a conference. Chapter 17, therefore, offers some guidelines on how to produce a clear and complete report of your work.

In the following chapters, all terms in **bold** are explained in the Glossary at the end of the book and the answers to end-of-chapter questions may be found at the end too. There are also a number of in-text questions (denoted by ■) designed to make you stop and think before you move on. Try covering the answer (denoted by ❏) and working it out for yourself before reading the suggested response.

PART I: TECHNIQUES
1 THE NIGHT SKY – POSITIONAL ASTRONOMY

In order to get the most out of the time you spend at an observatory, it is important to become familiar with the appearance of the night sky. Therefore it is instructive to consider how we measure the positions of celestial bodies and how their positions change during the night and during the year. For instance, if you go out at, say, 10 p.m. on a February evening, how does the clear night sky differ from that at 11 p.m. on the same night, or from that at 10 p.m. a week or so later? How do the Sun, the Moon and the planets move against the stellar background? Questions like these are important in relation to the whole of observational astronomy and planetary science, and you should be able to answer them after studying this section.

1.1 Celestial coordinates

To specify the position of a celestial body, we need a coordinate system that is fixed with respect to the distant stars, just as on Earth we have a latitude and longitude coordinate system that is fixed with respect to the Earth's surface. Indeed, this terrestrial system is the basis of a celestial system that is a natural choice for Earth-based observers.

The terrestrial coordinate system is shown in Figure 1.1. The position of any point O on the Earth's surface is specified by two angles, the latitude *lat* and the longitude *long*. Anywhere on the line l_O has the same latitude as O (this is a line of latitude) and anywhere on the line L_O has the same longitude as O (this is a line of longitude). Longitudes are measured from a zero of longitude that has to be chosen. For historical reasons, the zero is the line of longitude that passes through a particular telescope at the Royal Observatory, Greenwich, London. (This is often referred to as the Greenwich meridian.) Longitudes extend from 0 degrees of arc (deg, or °) to 180° east of Greenwich, and from 0° to 180° west of Greenwich. (Obviously 180° east and 180° west are coincident.) The zero of latitude is the Equator. This is the line on the Earth's surface midway between the North and South Poles. These poles are where the Earth's rotation axis meets the Earth's surface. Latitudes extend from 0° to 90° north of the Equator at the North Pole, and from 0° to 90° south of the Equator at the South Pole.

The degree of arc (1°) can be subdivided into 60 minutes of arc (60 arcmin, or 60′), and the minute of arc can be subdivided into 60 seconds of arc (60 arcsec, or 60″). To put these angular sizes into perspective, it is useful to remember that the angular diameter of the Sun and Moon, as viewed from the Earth, are both ≈ 30 arcmin = 0.5°, the apparent diameter of Jupiter is ≈ 45 arcsec when closest to the Earth, the fuzzy image of a star affected by turbulence in the Earth's atmosphere is ≈ 1 arcsec across, and the image of a star obtained by the Hubble Space Telescope, limited by the telescope's optics, is ≈ 0.1 arcsec across.

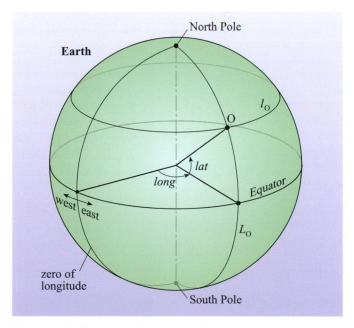

Figure 1.1 Terrestrial latitude and longitude.

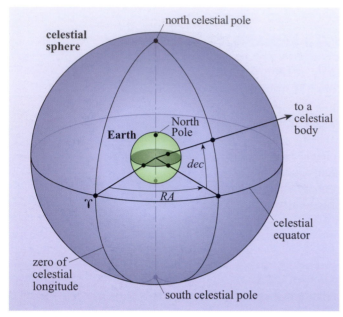

Figure 1.2   The celestial sphere.

To obtain a useful system of celestial coordinates, imagine the Earth to be surrounded by a sphere with the same centre as the Earth, as in Figure 1.2. This is called the **celestial sphere**. The radius of the celestial sphere is arbitrary, in other words, it can be made as large as we like. The line from the centre of the Earth through the North Pole intersects the celestial sphere at a point called the **north celestial pole**. The **south celestial pole** is defined in an analogous way. The projection of the Earth's Equator from the centre of the Earth on to the celestial sphere is called the **celestial equator**.

The Earth's rotation axis is very nearly fixed with respect to the distant stars. Certainly, on a human timescale, the north and south celestial poles and the celestial equator can be regarded as very nearly fixed. Therefore until further notice we shall regard them to be so. We thus have the basis of a useful coordinate system, the *celestial coordinate system,* used by astronomers and by navigators.

Suppose that there is a celestial body in the direction shown in Figure 1.2. Its celestial coordinates are the two angles *dec* and *RA*, celestial equivalents of latitude and longitude, respectively.

Celestial latitude is called **declination**, abbreviated as *dec* in Figure 1.2, and sometimes written as δ (the Greek lower case letter *delta*). Like terrestrial latitude, it extends from 0° at the celestial equator to 90° N at the north celestial pole, and to 90° S at the south celestial pole.

However, the usual convention with *celestial* latitude is to write northern latitudes as positive, and southern latitudes as negative. Each degree of declination is subdivided into 60 minutes of arc, and each minute of declination is subdivided into 60 seconds of arc.

Celestial longitude is called **right ascension**, abbreviated as *RA* in Figure 1.2, and sometimes written as α (the Greek lower case letter *alpha*). Note that right ascension is measured only eastwards from the zero of celestial longitude, and thus, in terms of degrees, runs from 0° to 360°. This is *one* difference from terrestrial longitude, which by convention, is measured from 0° to 180° east and from 0° to 180° west. There are two more differences. First, the zero of celestial longitude is *not* the projection onto the celestial sphere of the zero of terrestrial longitude (the Greenwich meridian).

■   Why would such a projection not be useful?

❑   Such a projected longitude would sweep across the stars as the Earth rotates, and so the celestial longitude of every celestial body would continuously change.

Therefore, the zero of celestial longitude is specified by a particular point on the celestial equator that is fixed with respect to the distant stars. It follows that the *RA* of a celestial body is also fixed. For historical reasons, the chosen point is called the **First Point of Aries**, which we shall denote by the symbol ♈ (the astrological symbol for the constellation Aries).

The second difference between terrestrial longitude and right ascension is that right ascension is not usually measured in degrees, but in hours! Right ascension ranges from 0 hours (+0° longitude) to 24 hours (+360° longitude), which is back at zero. This convention arises from the rotation of the Earth, which rotates once a day at the centre of the celestial sphere. As 360°/24 h = 15° h⁻¹, so 1 hour of right ascension corresponds to 15° at the celestial equator. The abbreviation for hours is h, and the usual subdivisions into minutes (min) and seconds (s) apply. Note that 1 minute of right ascension is not the same angular size as 1 minute of arc; at the celestial equator the former is one-sixtieth (1/60) of 15°, while the latter is one-sixtieth of 1° of arc. Similarly 1 second of right ascension at the celestial equator is 15 times larger than 1 arcsec.

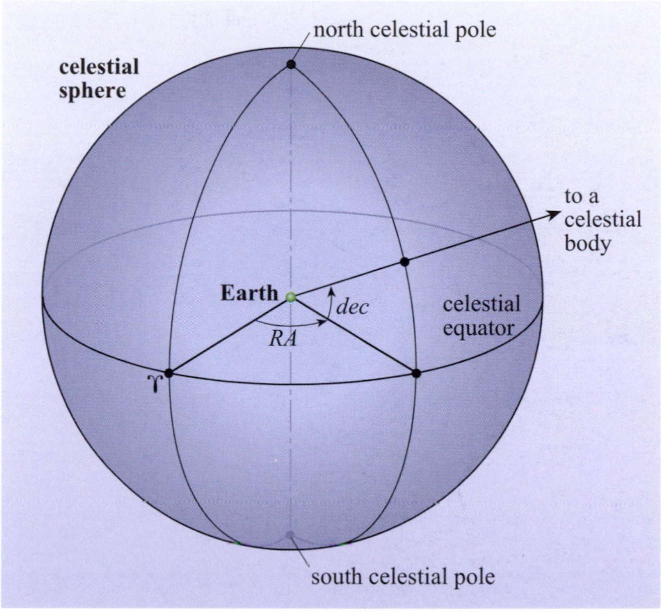

Figure 1.3   A large celestial sphere with the Earth shrunk to a small dot at the centre.

Strictly speaking, celestial coordinates are defined with respect to the Earth's centre. However, we can make the celestial sphere so much larger than the Earth that, for practical purposes, from any point on the Earth's surface the celestial sphere looks the same, and the celestial coordinates of distant celestial bodies have values practically independent of the position of the point on the Earth's surface. Such a celestial sphere is shown in Figure 1.3, where the Earth is so much smaller that it is shown as a point.

On the celestial sphere, note that the lines marking different right ascensions converge as they progress from the celestial equator to the celestial poles, in the same way that lines of longitude converge at the Earth's geographical poles. What this means is that although 1 minute of right ascension corresponds to 15′ (arcmin) at the celestial equator, it corresponds to fewer arcminutes at declinations away from the celestial equator. The factor by which it is reduced is simply the cosine of the declination, $\cos \delta$.

■ According to a particular catalogue, the star Vega has coordinates RA = 18h 36min 56.3s, dec = +38° 47′ 01″. Why do you think the right ascension has been specified to the first decimal place of seconds, while the declination is rounded to the nearest whole second of arc?

❑ At the celestial equator, 1 second of right ascension is 15 times larger than 1″. At the declination of Vega, the multiple is reduced by a factor: $\cos(38° 47′ 01″) = 0.78$, so 1 second of right ascension is 0.78×15 ~ 12 times larger than 1″ of declination. By writing the declination to the nearest arcsecond, the implied accuracy of the position is ±0.5″ in this coordinate. If the right ascension is known to the same accuracy, this corresponds to ±0.5″/(12″ s⁻¹) ~ ±0.04 seconds of right ascension at this declination. Hence if both coordinates are measured to the same accuracy, it makes sense to quote the right ascension to the nearest 0.1 seconds.

1.2 A map of the sky

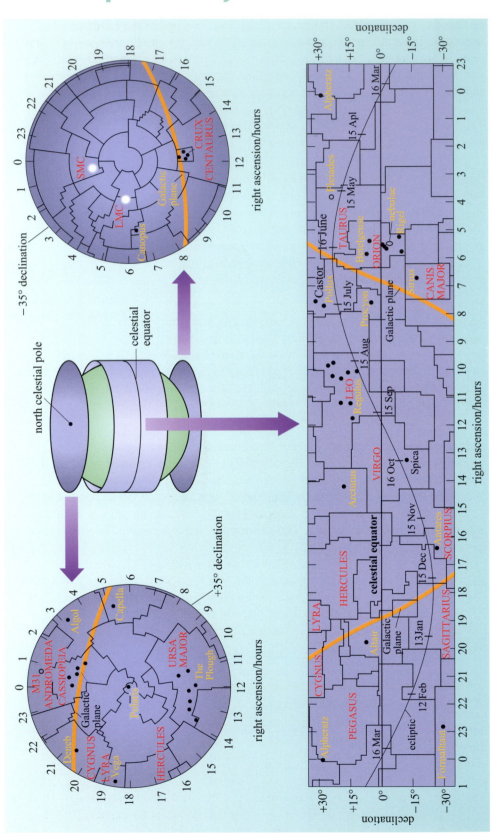

Figure 1.4 A map of the sky obtained by projecting the celestial sphere on to a flat surface consisting of three separate parts. The constellation boundaries and some stars are shown.

A sphere is not a very useful form of map for the printed page, and Figure 1.4 shows one way of projecting the celestial sphere on to a plane. We end up with a flattened view exactly as in Mercator projections of maps of the Earth, looking outwards. The entire sky (flattened or not) is divided up into 88 irregularly shaped regions, leaving no gaps. Each region is called a **constellation**, and all of them have names, though in Figure 1.4 only a few of the names are shown. These regions of sky are based around patterns of stars (also known as constellations), some of which are easily recognizable: here we have shown just the more prominent stars. Typical constellations are 10° to 20° across, as shown in Figure 1.5 which is a photograph of the familiar constellation Orion.

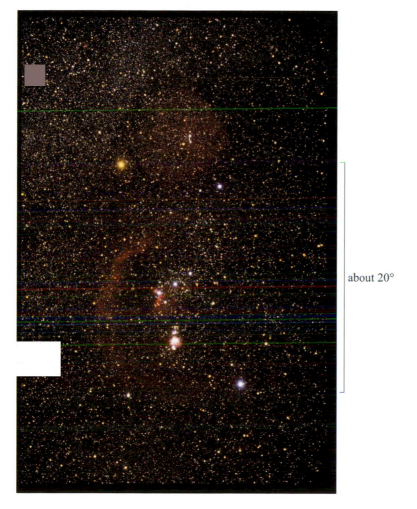

about 20°

Figure 1.5 A photograph of the familiar constellation Orion. The seven bright stars that locate Orion's shoulders, knees and belt are clearly seen, as is the bright patch known as the Orion nebula which is in Orion's sword. (© *Till Credner, AlltheSky.com*)

Any body in the sky will have celestial coordinates that lie in a particular constellation, and the way to express this is to say that object X lies *in* Orion, for instance. Indeed, this provides us with a basis for labelling the brighter stars. This is done by using the letters of the Greek alphabet (α, β, γ, δ, and so on), followed by the constellation name, which is usually reduced to a standard three-letter abbreviation. Thus, in Orion we have α Orionis, β Orionis, etc. (the modified constellation name is the Latin genitive), which are abbreviated to α Ori, β Ori, etc. To include more stars, numbers are used once the Greek letters are exhausted. The brightest stars also have individual names: thus α Ori is Betelgeuse ('betel-jers'), and β Ori is Rigel ('rye-jel'), for instance.

In order to include faint stars, and non-stellar objects, the constellation-based system is abandoned, and various catalogues are used. We shall not go into details, though you will come across a number of particular examples, notably the Messier Catalogue, and the New General Catalogue (NGC). These are both used for non-stellar objects and star clusters. The Messier Catalogue was prepared originally by the French astronomer Charles Messier (1730–1817), and today consists of 110 bright objects of assorted types, labelled M1 to M110. The New General Catalogue is much larger, embracing 7840 objects, including all those in the Messier Catalogue. It was published in 1888 by the Danish astronomer John Louis Emil Dreyer (1852–1926), who spent much of his working life in Ireland. Entries in his catalogue take the form NGC, followed by a number.

Modern astronomical catalogues contain vast numbers of stars and galaxies. For instance, the Hubble Space Telescope (HST) Guide Star Catalog contains coordinates and brightnesses of over 15 million stars; the catalogue resulting from the two-micron all-sky survey (2MASS) contains over 470 million objects and the USNO-B1.0 catalogue, constructed from scanning photographic sky survey plates obtained over the last fifty years, contains over 1 billion objects. The HST Guide Star Catalog is in common use in computer-based sky maps.

Some objects do not have constellation-based names, because their positions are not fixed and their celestial coordinates therefore move from one constellation to another. (In principle, all celestial bodies move with respect to each other and the positions of stars on the celestial sphere do change, but these changes are generally very small.) Notable amongst such bodies are the Sun, the Moon, and the planets, none of which is shown in Figure 1.4 for this very reason, though we have shown the annual path of the Sun – this is the thin sinusoidal line about the celestial equator labelled 'ecliptic'.

Before we leave Figure 1.4, note the star Polaris. With a declination of just over +89°, the position of this relatively bright star is currently close to the north celestial pole. Thus, it is often called the *Pole star*. The Southern Hemisphere is not as fortunately endowed: there are no relatively bright stars located this close to the south celestial pole.

1.3 The instantaneous view

Let's consider now the view of the sky that we get *at a particular instant* from a particular point on the Earth's surface, such as from the point O in Figure 1.6. The vertical direction at O is from the centre of the Earth upwards through O, and it is perpendicular to the horizontal plane at O. In this plane we have marked the north, east, south and west directions.

■ How much of the sky can an observer at O see?

❏ An observer at O can see only the half of the sky that lies above the horizontal plane.

The horizontal plane at O can be expanded to meet the celestial sphere much larger than the Earth, as in Figure 1.7. The plane intersects the sphere at the observer's horizon in the stars, dividing the sphere into the visible and invisible hemispheres. The vertical direction at O intersects the celestial sphere at a point called the **zenith**; this is the point overhead at O. The **north point** at O is where the north direction in the horizontal plane at O (Figure 1.6) intersects the celestial sphere, and likewise for the east, south and west points. The **meridian** at O is the arc that connects the north point, the zenith, and the **south point**; it also passes through the celestial poles: only the visible half of the meridian is shown in Figure 1.7.

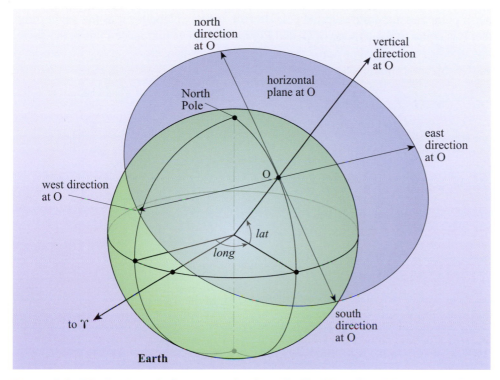

Figure 1.6   The horizontal plane at a point O on the Earth's surface. The three straight lines through O intersect each other at 90°.

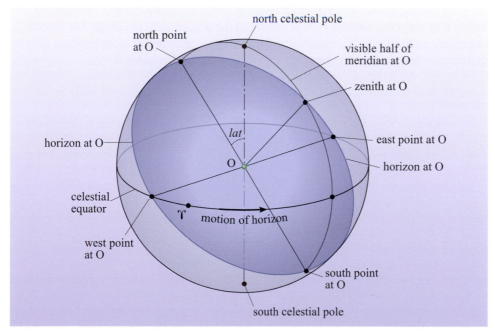

Figure 1.7   The horizontal plane at the point O on the Earth's surface (Figure 1.6), extended to meet a large celestial sphere.

The meridian is an example of a **great circle**. This is any line on a sphere that is the intersection between the sphere and a plane passing through its centre. The celestial equator is another great circle. When we measure angles between two points on the celestial sphere, we do so along the great circle that connects them. One example is along the meridian itself: in Figure 1.7 we have labelled the angle at O between the north point and the north celestial pole. In fact, we have labelled it as *lat*, suggesting that it is equal to the latitude of O, and Figure 1.8 shows that indeed this is the case.

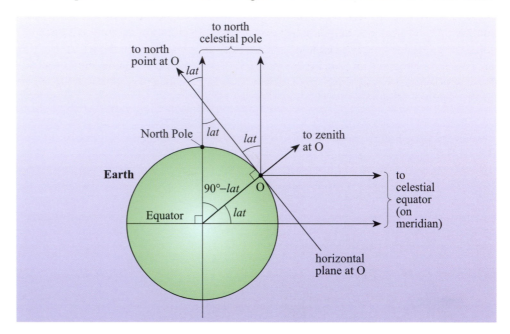

Figure 1.8 The angle at O between the north point and the north celestial pole is the latitude *lat* of O.

Often, the zenith angle is called the zenith distance.

Finally, we need to define the **altitude** and **zenith angle** of a celestial body. The altitude is the angle *alt* in Figure 1.9: it is the angle between the horizontal plane and the direction to the body, measured along the great circle that passes through the zenith and the point on the celestial sphere in the direction to the body. The zenith angle (*zen* in Figure 1.9) is simply the angle on the same great circle between the direction to the body and the zenith. Clearly we have *zen* + *alt* = 90°.

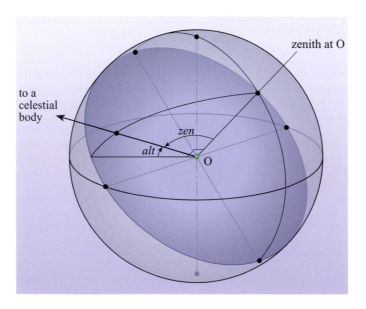

Figure 1.9 The altitude *alt* and zenith angle *zen* of a body.

1.4 Time and the effect of the Earth's motions

The visible hemisphere in Figure 1.7 for an observer at O on the Earth's surface is *not* fixed on the celestial sphere. This is because the Earth rotates on its axis once every day, anticlockwise as viewed from above the North Pole – this direction of rotation is called **prograde** rotation, because the planet spins in the same sense as the direction of orbital motion. A planet spinning in the opposite sense to its orbital motion is said to have **retrograde** rotation. The axis of rotation is along the line joining the north and south celestial poles. Thus, stars appear to make a circle around the north celestial pole once a day, with celestial bodies that are sufficiently far from the poles rising above the horizon in the east, and setting below the horizon in the west (see Figure 1.10).

Figure 1.10 A long exposure of the sky showing the apparent motion of the stars as the Earth rotates. This is in fact an image of the south celestial pole. (Courtesy of the *Anglo–Australian Observatory/David Malin Images*)

The Earth not only rotates on its axis, but also orbits the Sun, taking a year to return to the same position with respect to the distant stars, as seen from the Sun. The Earth revolves about the Sun in an orbit that, for present purposes, we can approximate as a circle with the Sun at the centre, and with the Earth moving at a uniform speed around the circle. The radius of the circle is 1.50×10^8 km.

■ What is the speed of the Earth in its orbit around the Sun?

❑ Speed = distance / time. The distance travelled by the Earth in one year is the circumference of a circle whose radius is 1.50×10^8 km. This distance is given by $2\pi \times$ radius $= 2\pi \times 1.50 \times 10^8$ km $= 9.42 \times 10^8$ km. So the speed of the Earth's motion in its orbit is 9.42×10^8 km/(365.26 day $\times$ 24 h day^{-1} $\times$ 60 min h^{-1} $\times$ 60 s min^{-1}) = 29.9 km s^{-1}.

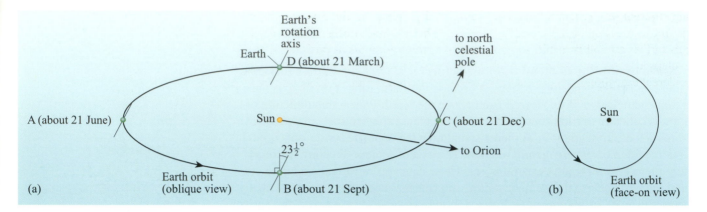

Figure 1.11 (a) An *oblique* view of the Earth's orbit around the Sun, showing the Earth's rotation axis. (b) Face-on, the Earth's orbit is far more circular.

The plane of the Earth's orbit is called the **ecliptic plane**. The Earth's axis of rotation is *not* perpendicular to this plane, but is inclined to the perpendicular at an angle of 23° 27′ as Figure 1.11 shows. Note that in Figure 1.11a the non-circular shape of the Earth's orbit arises from the oblique viewpoint: it appears much more circular when we view it face-on, as Figure 1.11b shows. Note also that the dots representing the Sun and the Earth are not to scale – they are far too large! Most important of all, note that the direction of the Earth's axis remains fixed (almost) with respect to the distant stars, and *not* with respect to the Sun. At A the Northern Hemisphere is maximally tilted towards the Sun. This is called the **June solstice**, and happens on or near 21 June each year. (In the Northern Hemisphere, this is more commonly referred to as the *summer solstice*. However we use the term June solstice to avoid confusion, since this is known as the *winter solstice* in the Southern Hemisphere!) The Earth moves anticlockwise around the Sun, as viewed from above the northern side of the ecliptic plane – prograde motion again. Thus, one-quarter of an orbit later we reach B around 21 September, then C around 21 December (the **December solstice**, also known as the *winter solstice* in the Northern Hemisphere and the *summer solstice* in the Southern Hemisphere), and then D around 21 March.

Figure 1.11 demonstrates, through the example of the constellation Orion, that certain celestial objects are visible from the Earth at only certain times of the year. This lies in the direction shown. Therefore, around May and June each year, as seen from the Earth, Orion lies in about the same direction as the Sun, and so will not be visible – it is only above the horizon during the daytime and the sky will be too bright. By contrast, around November and December, it lies in roughly the opposite direction from the Sun, and so will be visible for most of the night. During the other months it will be visible for less of the night. A different case is that of Polaris. The proximity of its direction to the north celestial pole means that in the Northern Hemisphere, Polaris never sets, so it is visible during the hours of darkness.

■ What is the visibility of Polaris in the Southern Hemisphere?

❏ It never rises, so it is never visible.

To examine further the changing relationship between the Sun and stars, let's examine the Earth at position 1 in Figure 1.12, where the direction to the Sun is on the observer's meridian – this defines local *noon*. Also, suppose that a star lies in the direction of this meridian. (There needn't be a real star, any point fixed on the celestial sphere will do.) As the Earth moves around its orbit it also rotates, and at position 2 it has rotated just once with respect to the distant stars, and so the star again lies in the direction of the meridian. The time elapsed between positions 1 and 2 is one **sidereal day** ('sidereal' means 'star related'). However, the Sun is not yet again in the direction of the meridian. The Earth has to rotate farther (and it also moves farther around its orbit) to achieve this configuration, as at position 3. The time elapsed between positions 1 and 3 is one **solar day**, and it is clearly longer than the sidereal day, though by only a few minutes (note that Figure 1.12 is not drawn to scale).

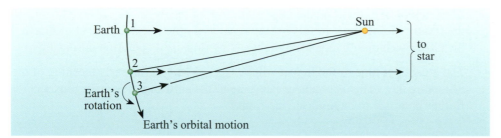

Figure 1.12 The difference between the sidereal day and the solar day: the straight black arrows denote the direction of the observer's meridian (not to scale).

The familiar 24 hour day of civil time is not *quite* the same as the solar day. This is because during the year there are small variations in the intervals between the Sun's crossing of an observer's meridian, and hence in the length of the solar day. These variations arise mainly from the tilt of the Earth's rotation axis, and from the Earth's orbit being slightly different from circular; we will not go into details. By contrast, the **mean solar day** is fixed in duration and equal to 24 hours. If solar time and mean solar time coincide at some instant, they will coincide again a year later, but in between differences develop, sometimes solar time being ahead of mean solar time, and sometimes behind. The civil day is equal in length to the mean solar day.

If we use the terms hours (h), minutes (min) and seconds (s) to denote subdivisions of the 24 h mean solar day (or civil day), then the sidereal day is 23 h 56 min 04 s long, i.e. 03 min 56 s shorter. This difference means that, according to mean solar time, the distant stars cross the observer's meridian 03 min 56 s earlier every day, which amounts to 27 min 32 s earlier every week. Thus, the stellar sky at say 22 h 00 min 00 s (mean solar time) on a certain day, looks the same as the stellar sky at 21 h 32 min 28 s seven days later. The stellar sky thus progresses westwards with respect to the Sun, taking one year for the differential cycle to be completed.

This extra rotation per year of the stellar sky means that there is one more sidereal day in the year than there are mean solar days. Thus, with 365.26 mean solar days per year, there are 366.26 sidereal days per year.

Mean solar time is a local time, in that observers at different longitudes will observe the Sun in the direction of their meridians at different times. For civil purposes this is very inconvenient, and so the world is divided up into time zones, where **civil time** is the same within each time zone. Mean solar time will vary with longitude across the zone, typically by one hour from the west to the east extremities of the zone. In most countries there is also **daylight saving time**, in which civil time is advanced by an hour in the spring and summer months: in the UK this is called British Summer Time.

When astronomers wish to record the time at which an observation is made, or at which a given phenomenon occurs, there is clearly scope for confusion due to the variation in civil time from one location to another. To get around this problem, astronomers therefore usually record the **universal time** (UT) associated with a particular event. To most intents and purposes UT is identical to the civil time on the Greenwich meridian, *without* the inclusion of daylight saving time. UT is measured in hours, starting from 0 at midnight, and incrementing by 24 hours over the course of one mean solar day.

■ An eclipse of a binary star is observed to occur at 22:15 hours civil time from an observatory in Chile. Civil time in Chile is 4 hours behind that at the Greenwich meridian. What is the UT of the event?

❑ The UT of the eclipse is 02:15 hours on the following day.

■ A transit of Io across the face of Jupiter is predicted to occur at UT 23:30 hours. What will be the local civil time of the event, as observed from an observatory in Majorca in July? (Majorcan civil time is 1 hour ahead of that at the Greenwich meridian, and daylight saving time is in operation during July.)

❑ The local civil time of the transit will be 01:30 hours in the early morning of the following day. (This is one hour later due to the shift in time zones, plus a further hour due to daylight saving time.)

When astronomers wish to be even more precise, they may use **coordinated universal time** (UTC). This is based on atomic time. To correct for the small irregular varying motion of the Earth around its axis, and in its orbit around the Sun, UTC is kept in step with UT by the insertion or deletion of leap seconds at the end of June or December as necessary.

1.5 The changing celestial coordinates of the Sun

Let's now add a celestial sphere to each Earth in Figure 1.11a. We obtain Figure 1.13. Note that as the Earth orbits the Sun the celestial sphere remains fixed with respect to the distant stars, and this is indicated by the fixed orientation of the line joining the north and south celestial poles, and by the fixed direction of Aries ♈, which lies on the zero of right ascension. The positions on the celestial sphere of even the nearer stars barely change as the Earth orbits the Sun, because the Earth's orbit is very small compared with their distance. However, the Sun's celestial coordinates change dramatically, and we shall now use Figure 1.13 to map these changes.

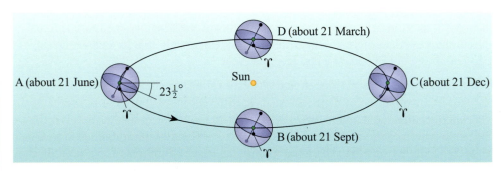

Figure 1.13   The changing celestial coordinates of the Sun.

When the Earth is at A the Sun has a right ascension 6h and a declination +23.5° (the June solstice). At B the values are (12h, 0°). The zero declination at B means that the direction to the Sun is on the celestial equator, with the declination changing from north to south. At such a time there is close to 12 hours between sunrise and sunset all over the Earth, as Figure 1.14 shows.

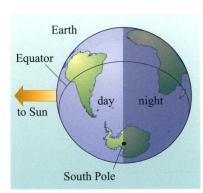

Figure 1.14   The Earth in position B in Figure 1.13, viewed from a point in its orbit just behind the Earth.

Such an event is loosely called an *equinox* (equal day and night) but strictly an equinox is the precise moment when the direction to the centre of the Sun is on the celestial equator. The position at B may be referred to as the **September equinox**, although in the Northern Hemisphere it is more commonly called the *autumnal equinox*. At C the Sun's celestial coordinates are (18h, −23.5°) (the December solstice). At D we have the other equinox, which may be referred to as the **March equinox**, with the Sun at (0h, 0°) and moving from south to north. Again note that this position is more commonly referred to as the *vernal equinox*, or *spring equinox*, by observers in the Northern Hemisphere. As with the solstices though, we designate the equinoxes using the month rather than the season to avoid confusion between Northern- and Southern-Hemisphere observers. The complete trace of the Sun's celestial coordinates is shown in Figure 1.15, and is called the **ecliptic** (also shown in Figure 1.4). Note that this is a great circle on the celestial sphere, the intersection of the ecliptic plane with the celestial sphere.

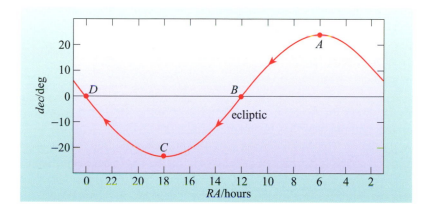

Figure 1.15 The changing celestial coordinates of the Sun. A is ~ 21 June, B ~ 21 September, C ~ 21 December, D ~ 21 March.

You have just seen that at the March equinox the *RA* of the Sun is zero, and so the Sun lies in the direction of ♈. *This is no coincidence*, for this is how the zero of right ascension is defined! Thus, the zero of celestial longitude is the line of celestial longitude that passes through the celestial equator at the point where the declination of the Sun is changing from south to north.

1.6 Precession of the Earth's rotation axis

If the Earth's rotation axis really was fixed with respect to the distant stars, then the direction to ♈ would be similarly fixed. Unfortunately, the axis is *not* quite fixed. Although this effect is small, it must be accounted for, since over time the coordinate system shifts relative to the positions of stars and hence their catalogue positions become incorrect. The largest effect by far is the so-called **precession** of the Earth's rotation axis, shown in Figure 1.16. The Earth's rotation axis takes 25 800 years to complete one circuit, so celestial coordinates go through a 25 800 year cycle, meaning they shift by up to 50″ per year. Consequently, a date (perhaps confusingly, also called the **equinox**) must always be appended to celestial coordinates to specify the time at which they are correct. It is customary to tabulate catalogue coordinates for the beginning of a year, so the equinox may be specified by the year only, e.g. 2000.0, rather than for every possible date. Hence the coordinates of the star Vega, for example, may be given as *RA* = 18h 36min 56.3s, *dec* = 38° 47′ 01″ (equinox 2000.0) or as *RA* = 18h 35min 14.7s, *dec* = 38° 44′ 10″ (equinox 1950.0).

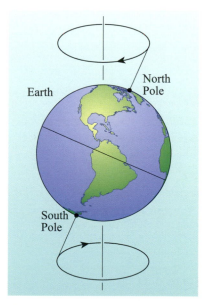

Figure 1.16 The precession of the Earth's rotation axis.

1.7 The apparent motion of the Moon and planets

The Moon orbits the Earth, and is easily the Earth's nearest neighbour, in fact nearly 400 times nearer than the Sun. Figure 1.17 shows the lunar orbit (enlarged) in relation to the Earth's orbit on four occasions during the year. The plane of the lunar orbit makes only a small angle (around 5°) with respect to the ecliptic plane. Therefore, as the Moon orbits the Earth, its celestial coordinates are never far from the ecliptic in Figure 1.15. Note that we have added no dates to Figure 1.17. This is because the orientation of the orbit of the Moon is not quite fixed with respect to the Earth's orbit.

Figure 1.17 The Moon's orbit (enlarged): an oblique view.

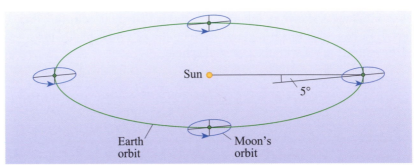

Figure 1.18 shows the Moon in its orbit; the size of the Moon is *greatly* exaggerated, and the orbit has been approximated by a circle, with the Earth at its centre. The Moon shines only by reflecting the Sun's radiation, and so the lunar phases depend only on the angle between the Moon and the Sun, as seen from the Earth. Thus, when the Moon and Sun lie in roughly the same direction we have a new Moon – essentially invisible unless the Moon passes exactly between an observer and the Sun, when they observe a *solar eclipse*. A new Moon occurs at intervals, on average, of 29.53 days. At the other extreme, when the Moon and Sun lie in roughly opposite directions, the Moon is full. When the angle between the Moon and the Sun is 90°, the Moon is half-full; these positions are marked *first quarter* and *third quarters* in Figure 1.18.

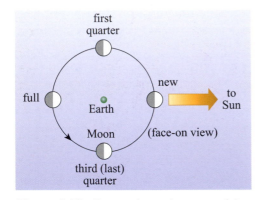

Figure 1.18 Lunar phases (not to scale).

■ In what sense are these positions quarters?

❏ They are one-quarter and three-quarters of the time through the lunar cycle, starting at new Moon.

The Moon rotates on its axis in the same time that it takes to orbit the Earth, and in the same direction. This is called **synchronous rotation**. As a result, the Moon presents more or less the same face towards the Earth – the familiar 'man in the Moon' pattern. However, largely because the Moon's orbit around the Earth is not perfectly circular, and also because its rotation axis is not quite perpendicular to the plane of its orbit, we do see rather more than half the Moon's surface. The Moon's surface, *as we see it*, appears to oscillate slightly in various ways around a mean position. The apparent oscillations are called **librations** and they allow us to see about 59% of the lunar surface from the Earth.

Beyond the Moon we come to the planets. These move in orbits around the Sun, in planes that, in most cases, make small angles i with respect to the ecliptic plane (Table 1.1). Therefore, the celestial coordinates of these planets stay close to the ecliptic. The exact positions vary from planet to planet, and from year to year.

Table 1.1 The planets and their inclinations i to the ecliptic plane.

Planet	i/deg
Mercury	7.00
Venus	3.39
Mars	1.85
Jupiter	1.30
Saturn	2.49
Uranus	0.77
Neptune	1.77
Pluto	17.2

1.8 Planispheres and planisphere software

We are now in a position to consider, in practical terms, the effect of the Earth's motions on an observer's view of the sky. In particular, we want to establish which half of the celestial sphere is above an observer's horizon at any particular date and time. A **planisphere** is a device that supplies this information, and an example is shown in Figure 1.19. When you are at an observatory, making observations, a planisphere can provide a quick and easy way of identifying what is visible from your location and when is the best time to observe a given object. Alternatively, there

Figure 1.19 A planisphere. Note that the zenith is half-way along the visible half of the meridian, i.e. almost coincident with the star Capella.

Figure 1.20 An image generated by a software planisphere package ('The Sky' Astronomy Software) showing the night sky from a particular location at a particular time.

are many different software packages (Figure 1.20) that effectively act as electronic planispheres (in addition to providing a lot more features as well). These software packages allow you to input the location of the observatory, and the date and time at which observations are to be made, and then return a detailed map of the sky for that particular instant. Whether a physical planisphere or a software planisphere is most useful for your observations will depend on your own preference and on what is available to you. Details of software packages will vary from one product to another, so it is not appropriate to give instructions for their operation here. However, all physical planispheres operate in exactly the same manner, so it *is* useful to provide some instruction in their use.

A planisphere is specific to a given latitude, give or take a few degrees. The lower sheet displays all of the sky ever visible from that latitude. Around the edge of the lower sheet, the right ascension is given in hours and in degrees, and along several of the radial lines the declination is given. The lower sheet also carries a scale that shows various dates through the year – we shall come back to this shortly.

The upper sheet rotates on the lower sheet around a point that represents the north celestial pole (we shall assume that the observer has a northern latitude). This sheet has an elliptical aperture, the boundary of which represents the observer's horizon.

Within the aperture is the hemisphere of sky on view at a particular date and time. To select a particular date and time, the desired local mean solar time on the upper sheet is lined up with the date on the lower sheet. Civil time should be sufficiently close to local mean solar time, except that, if daylight saving time is in force, then you should subtract an hour from civil time.

In Figure 1.19, we have marked the *visible half of the meridian*, the *zenith*, the *north celestial pole*, and the *north point* and *south point*. The planisphere itself marks the eastern and western horizons. (Note that, because we are representing a portion of a sphere on a plane, the planisphere has geometrical distortions.) Compared with terrestrial maps, you can see that east and west are reversed. This is because the planisphere is to be held above your head, and you are to look up at it, with the meridian correctly orientated. That's all there is to using it!

You can also use a planisphere to simulate the effect of the Earth's rotation. If the upper sheet is slowly rotated, in a clockwise direction, the stars rise in the east, reach their highest altitude as they cross the meridian, and set in the west. By lining up a particular time with a particular date, say 22 h (10 p.m.) on 14 February, and rotating the upper sheet clockwise until the same time lines up with 15 February, you have simulated the passage of one mean solar day. Note that, in such a case, the sky has shifted slightly westwards – this is because of the difference between the solar and sidereal days (due to the motion of the Earth about the Sun).

By rotating the upper sheet, the effect of declination on the apparent motion of a celestial body across the sky may also be demonstrated. At a sufficiently large negative (southerly) declination, a celestial body never rises, but remains always below the horizon. Then, as we select bodies with more and more northerly declinations, there comes a point at which a body just touches the horizon at the south point (the declination of this object is still likely to be less than zero). As the declination becomes yet more positive, the rising and setting points move farther away from the south point, and the time between rising and setting increases. The maximum altitude of the body, which occurs when it is on the meridian, also increases. We ultimately reach a declination above which a given body is *always* above the horizon. Such a body is called a *circumpolar* body. The higher the latitude, the greater the fraction of the sky that is circumpolar.

1.9 Summary of Chapter 1 and Questions

- Celestial coordinates are based on the *celestial sphere*. The celestial poles and the celestial equator are projections of the Poles and the Equator of the Earth.

- Celestial latitude is called *declination* (*dec*), and is measured in degrees north and south of the celestial equator. Degrees north are denoted by positive values, and degrees south by negative values.

- Celestial longitude is called *right ascension* (*RA*), and is measured eastwards in hours, where a 24-hour change is equivalent to a 360° change in celestial longitude. The zero of right ascension is marked by the *First Point of Aries*, ♈, defined as the point at which the Sun crosses the celestial equator moving from south to north. The north and south celestial poles, the celestial equator, and ♈ are all very nearly fixed with respect to the distant stars.

- At any instant, an observer on the Earth's surface will see only one-half of the celestial sphere.

- *Civil time* is based on the Sun, with the familiar day more accurately called the *mean solar day*. The *sidereal day*, based on the apparent positions of the stars, is about 3 minutes 56 seconds shorter than the mean solar day: any star will rise tomorrow about 3 minutes 56 seconds earlier than it did today.

- Times of astronomical phenomena are usually recorded in terms of their *universal time*. UT is essentially the same as civil time on the Greenwich meridian (GMT), ignoring the effect of *daylight saving time*.

- The Earth's orbital motion, and the inclination of its rotation axis with respect to its orbital plane (the ecliptic plane), cause the Sun's celestial coordinates to change. The Sun traces out a path on the celestial sphere called the *ecliptic*, which is the intersection of the ecliptic plane with the celestial sphere.

- The celestial coordinates of the Moon and the planets also change. The orbital planes of most of these bodies make small angles with respect to the ecliptic plane, so their celestial coordinates trace out paths on the celestial sphere close to the ecliptic.

- A *planisphere* is a useful tool for displaying the celestial hemisphere visible to an observer at a given latitude, at any time of the day and on any date in the year. Software tools can do this also, and usually more besides.

- As the Earth rotates, the visible hemisphere changes, but (except for an observer at the Earth's Equator) there is a portion of sky that never sets, and another portion that never rises. These portions depend solely on the observer's latitude.

QUESTION 1.1

The celestial coordinates of the star β Trianguli are (02h 09.2min, +34° 57′), and those of the star θ Centauri are (14h 06.4min, −36° 20′). Roughly speaking, how far apart (in angular terms) are the directions of the two stars in the sky?

QUESTION 1.2

(a) What is the altitude of (i) the north celestial pole (for an observer in the Northern Hemisphere) and (ii) the zenith?

(b) Suppose that, from some point on the Earth's surface, the Sun, at some particular moment, has an altitude of 64° 21′. How many degrees and minutes of arc is it from the zenith?

QUESTION 1.3

For observers at latitudes 90° N, 50° N and 0°, describe, with the aid of sketches a bit like Figure 1.8, where the celestial equator lies in the sky in each case.

QUESTION 1.4

Refer to the image of the planisphere in Figure 1.19.

(a) What civil time on 27 December corresponds to the view of the sky shown?

(b) Which constellations are just setting at this time?

(c) Roughly what is the right ascension of stars that lie on the southern part of the meridian at this time?

(d) Roughly what is the most southerly declination that is visible at this time?

2 TELESCOPES

Unaided human eyes, well as they may serve the needs of everyday life, are not very suitable for detailed astronomical observation. First, the eye has a limited sensitivity. A distant source of light, such as a star, will not be seen at all unless the intensity of light from it reaching your eye is above the sensitivity threshold of the retina. Second, the ability of the eye to distinguish fine detail is limited by the finite physical size of the detectors on the retina and by the small aperture of the eye. The ability of a telescope or the eye to distinguish between two objects that are very close to one another is called its (angular) resolution. Limited resolution makes it impossible for human eyes to separate individual distant sources of light that are closer than about 1′ apart, or to discern details of their shape or structure on angular scales finer than this.

The invention of the telescope at the beginning of the seventeenth century was an important milestone in the advancement of astronomy. Here was a simple instrument that at once overcame, to some degree at least, these shortcomings of human eyes. In this section we shall first look at the characteristics of optical elements that may be combined to make telescopes, then we shall consider the main designs of **refracting telescopes** and **reflecting telescopes** that have been developed over the past four centuries, and finally look at the key ways of characterizing the performance of an astronomical telescope. To clarify the nomenclature, the name 'refracting telescope' (or refractor, for short) is used to indicate a telescope in which only lenses are used to form the image; the name 'reflecting telescope' (or reflector for short) is used to indicate a telescope in which a curved mirror is used in place of one of the lenses. In any telescope, the optical element that gathers the incoming light is variously referred to as either the **objective** lens or mirror, or the **primary** lens or mirror. A subsequent **lens** used to view the image by eye is referred to as the **eyepiece lens**.

2.1 Optical elements

In order to understand how telescopes work, it is useful to outline the basic principles of curved lenses and mirrors. A surface which is the same shape as a small portion of a sphere is called a spherical (or more correctly spheroidal) surface. Surfaces with this shape have a special optical property which makes them highly valuable: their ability to bring light to a focus. Actually, the focusing properties of a spheroidal surface are not perfect, as we shall see later, but the imperfection is often more than compensated for by the purely practical consideration that a precise spheroidal optical surface can be produced much more easily – and hence at much lower cost – than a precise aspheroidal (non-spheroidal) optical surface.

Three important focusing properties of spheroidal surfaces are described in the three following statements. Unfortunately, neither of the first two statements is exactly true for any real optics, but they are extremely valuable approximations to the truth and will greatly aid your ability to understand the layouts of optical instruments such as telescopes and spectrographs.

(1) When parallel rays of light pass through a lens with convex spheroidal surfaces, or reflect from the surface of a spheroidal concave mirror, they are brought to a focus. The distance of the **focal point** from the lens (or mirror) is called the **focal length**, f. This is a single quantity that characterizes the optical performance of the lens or mirror in question.

(2) Light rays passing through the centre of a lens do not deviate from their original path.

(3) Light paths do not depend on the direction in which light is travelling. So, for example, since parallel rays of light are brought to a focus by a convex lens at a distance f from the lens, then rays of light emanating from a point a distance f away from the lens will be converted into a parallel beam. A lens which is used in such a way is called a **collimator**, and the beam of parallel light that is produced is said to be **collimated**.

Broadly speaking there are two sorts of lenses and mirrors used in optical systems. Converging (convex) lenses and converging (concave) mirrors each cause parallel rays of light to come together at the focal point, or focus, of the lens or mirror (Figure 2.1a and b). In contrast, diverging (concave) lenses and diverging (convex) mirrors each cause parallel rays of light to spread out as if emanating from the focal point situated at a distance of one focal length from the centre of the lens or mirror concerned (Figure 2.1c and d).

Converging lenses and mirrors used individually can each produce **real images** of distant objects, by which is meant an image that may be captured on a screen or directly on a detector such as photographic film. Real images are those images made by the convergence of actual rays of light. However, when eyepiece lenses are used with telescopes, the final image formed by the telescope is said to be a **virtual image**, since it is situated at a location from which rays of light appear to emanate (see Figure 2.2 and Figure 2.3 below). Such an image cannot be captured directly on a detector. However, eyepieces are always used in conjunction with another lens – namely the lens of the eye itself – which converts the virtual image produced by the telescope into a real image on the retina of the eye.

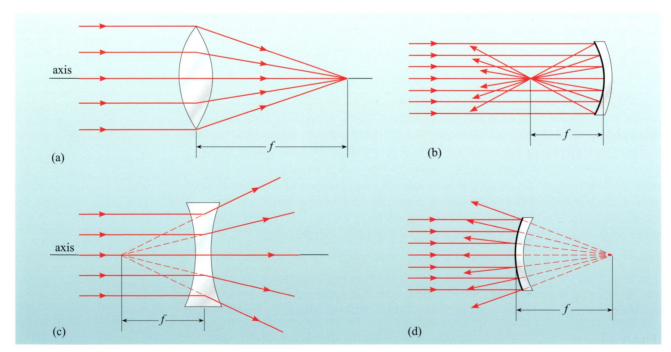

Figure 2.1 (a) A convex lens will cause parallel rays of light to converge to the focal point. (b) A concave mirror will cause parallel rays of light to converge to the focal point. (c) A concave lens will cause parallel rays to diverge as if from the focal point. (d) A convex mirror will cause parallel rays to diverge as if from the focal point. The reflecting surface of the mirror is shown by a thicker black line.

Two additional comments should be made relating to the term 'focal length'. Firstly, a series of two or more lenses and/or mirrors can also bring parallel incident light rays to a focus, though obviously at a different point from that of any of the elements independently. The focal length of such a series of optical elements is defined as the focal length of a single lens that would bring the same rays of light to a focus at the same angle of convergence. The *effective* focal length may therefore be quite different from the actual distance between the optics and the focus. As we shall see later, this allows long focal lengths to be compressed into short path lengths.

Secondly, it is sometimes common to quote the number that is obtained by dividing the focal length of an optical assembly by the diameter of the bundle of parallel light rays that is brought to a focus. In some optical systems, such as telescopes, the diameter of this bundle of light rays is the same as the diameter of the main optical element, though this is not always the case, particularly for most camera lenses. The number obtained by calculating this ratio is referred to as the **f-number**, written f/# or F/# where # is the numerical value.

■ What is the f-number of a 200 mm diameter telescope with a focal length of 2400 mm?

❏ The f-number is 2400 mm/200 mm = 12. This would be written f/12 or F/12.

2.2 Refracting telescopes

The story of telescopes began in 1608, when a Dutch optician, Hans Lippershey, discovered that a distant object appeared larger when viewed through a combination of two lenses: a relatively weak (i.e. long focal length) converging lens facing the object and a strong (i.e. short focal length) diverging lens in front of the eye. This combination of lenses was subsequently used by Galileo Galilei for looking at the Moon, the planets and the stars, and it became known as the **Galilean telescope** (see Figure 2.2).

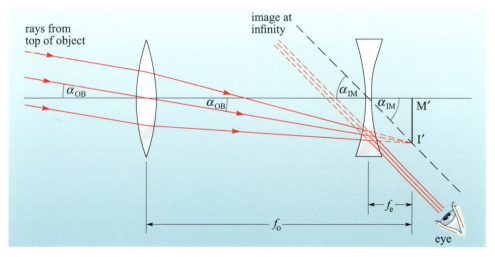

Figure 2.2   A Galilean (refracting) telescope. Parallel rays of light from a distant object would be brought to a focus in the focal plane of the (converging) objective lens. However, the (diverging) eyepiece lens intercepts these rays and renders them parallel once more, but travelling at a larger angle to the optical axis. This leads to an increase in the apparent angular size (i.e. the image is magnified with respect to the object). The final image is a virtual image, located at infinity and is the same way up as the object.

By about 1630 Johannes Kepler had replaced the diverging eyepiece lens with a converging lens of very short focal length. This new combination of two converging lenses, the **Keplerian telescope**, has remained the principal form of construction of refracting astronomical telescopes until this day, although many technological improvements have been introduced to cope with the various problems that set limits on the basic telescope's performance. Figure 2.3 shows a diagram of a refracting telescope of this type.

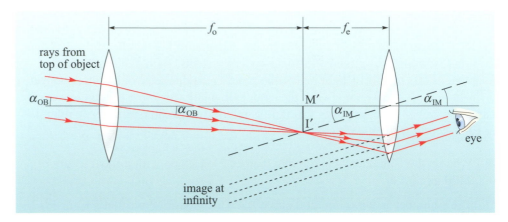

Figure 2.3 A Keplerian (refracting) telescope. Parallel rays of light from a distant object are brought to a focus by the (converging) objective lens and then diverge as they approach the eyepiece lens. This converging lens renders the rays parallel, but travelling at a larger angle to the optical axis. As in the Galilean telescope the virtual image is therefore magnified with respect to the object, and is located at infinity. This image is inverted.

To optimize the light-gathering power of an optical telescope, the aperture D_o of its objective lens must be as large as possible. Unfortunately, this is easier said than done. To begin with, there are serious technological problems in producing very large lenses. To ensure that the initial block of glass, from which the lens is to be made, is perfectly transparent and optically homogeneous throughout, the molten glass may need several years (!) of gradual and controlled cooling. Next comes the problem of grinding and polishing – it is not easy to sustain a perfect spherical curvature for a very large focal length lens over the whole of its surface area. And when you have a large lens, it is inevitably a thick lens, which therefore absorbs light, preferentially in the blue and violet part of the spectrum. It is also a very heavy lens, which means that it would have a tendency to sag under its own weight. In practice, usable objective lenses with a diameter much larger than 1 metre cannot be made. Figure 2.4 shows a photograph of one of the largest refracting telescopes in the world, the 36 inch refractor at the Lick Observatory, California. Note the extremely long body of the telescope in relation to its diameter.

Achieving high magnification with a telescope requires a long focal length f_o, but limits on the maximum possible value of f_o are set by the need to make the whole instrument movable. It is clear from Figure 2.3 that the physical length of a Keplerian refracting telescope cannot be less than f_o. Hence, it would hardly be realistic to plan a telescope with a focal length of 100 metres using this design! However, it is important to remember that achieving high magnification is not necessarily always useful, and sometimes it is better to have very *short* focal lengths. This will increase the field-of-view of the telescope and make the images appear brighter, as the light is less spread out. Designing optics with very short focal lengths leads to some optical aberrations, which we discuss briefly.

Figure 2.4 The 36 inch refractor at the Lick Observatory, California. (© *UCO/Lick Observatory*)

Optical aberrations are not errors of manufacture, but are undesirable physical characteristics of refracting and reflecting surfaces. For example, parallel rays of light passing through different parts of a lens are not focused to the same point by spherical surfaces; this is known as **spherical aberration**. This wouldn't be a problem except for the fact that spherical surfaces are relatively easy to produce, whereas parabolic surfaces, which give a perfect focus, are much more difficult to produce. Even from the same part of the lens though, waves of different frequency (i.e. colour) are focused to different points; this is known as **chromatic aberration**. By combining several lenses of different optical strengths and different materials, chromatic aberration can be reduced, but the problems are formidable and increase with the increasing size of the lenses and with the angle of the rays with respect to the optical axis. Thus, in practice, refracting telescopes have only a relatively narrow field-of-view within which the resolution is good.

Two other types of aberration that frequently affect images that lie off the optical axis are *coma* and *astigmatism*. Coma arises because each annular zone of the lens or mirror produces an off-axis image of a point source of light (or star) in the form of a circular patch of light. These circles vary in position and diameter moving from zone to zone, so that the combined 'point-image' in the focal plane is a fan-shaped area formed from overlapping circles. Astigmatism occurs because light that falls obliquely on a lens or mirror is focused not as a single point, but as two perpendicular lines, each at different distances from the lens or mirror. At the best focus position, the image of a point source will appear as an elliptical shape.

The net result of all these problems is that large refracting telescopes are no longer built for serious astronomical work.

2.3 Reflecting telescopes

A lens is not the only object that can collect and focus light and thus produce visual images. People have known about and used mirrors for much of recorded history, but it took no less a genius than Isaac Newton to realize how a curved *mirror* could be used to construct an optical telescope, and that this would overcome some of the most important shortcomings of refracting telescopes.

As noted earlier, a concave spherical mirror will reflect parallel rays approaching along its axis of symmetry so that they come together almost at one point (the focus) lying between the reflecting surface and its centre of curvature. The main advantage of focusing by reflection is that the angle of reflection is the same for all wavelengths in the incident radiation. So there is no analogy to the chromatic aberration that takes place in lenses. Hence, if we replace the objective lens of a telescope with a reflecting spherical mirror, we have automatically and completely eliminated the chromatic aberration on the input side of the telescope (we still have it in the eyepiece). However, there is still spherical aberration because rays reflected from the points further away from the axis of symmetry will be focused nearer to the reflecting surface, as shown in Figure 2.5.

The difference of focus shown in Figure 2.5 is exaggerated, to make the point clear. However, the spherical aberration of a converging mirror is always less than the spherical aberration of a converging lens of the same focal length. For converging mirrors that are only small parts of the hemisphere, it can usually be neglected. Unfortunately, by reducing the size of the mirror to reduce spherical aberration, some of the potential light-gathering power is lost, and the useful field-of-view is

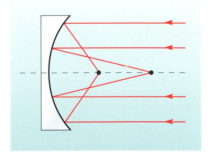

Figure 2.5 Spherical aberration of a concave mirror (exaggerated for clarity). The point to which parallel rays of light are focused depends on the distance of the incident rays from the optical axis. Incident rays initially far from the optical axis are brought to a focus nearer to the mirror surface than rays travelling close to the optical axis.

also limited. Fortunately, there are two ways of dealing with this problem. We can either choose a paraboloidal shape for the mirror (as in Figure 2.6) or we can correct the focusing of a spherical mirror by introducing a suitable pre-distortion into the incoming wavefront. This is done by placing in front of the mirror a transparent plate of such a shape that it refracts the initially parallel rays near the optical axis differently from those further away from it (as shown in Figure 2.7). This correcting plate is known as a **Schmidt plate**, and the reflecting telescopes in which a Schmidt plate is used are called **Schmidt telescopes**.

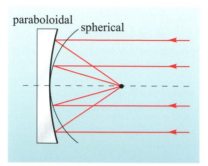

Figure 2.6   The elimination of spherical aberration using a mirror of paraboloidal shape. Parallel rays of light are all brought to the same focus, irrespective of their distance from the optical axis.

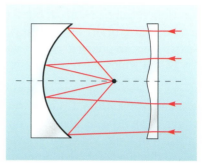

Figure 2.7   The Schmidt correcting plate for compensating for spherical aberration. Parallel rays of light further from the optical axis are bent with respect to rays closer to the optical axis, by the Schmidt plate. The net result is that all rays are brought to a common focus.

In case you are wondering how you could actually see the image of a star produced by a spherical converging mirror without being in the way of the oncoming light, this problem was solved simply and neatly by Newton as shown in Figure 2.8. He put a small flat mirror (the **secondary** mirror) just before the focus of the primary mirror and at an angle of 45° to the optical axis. He thus moved the image towards the side wall of the telescope tube, where he then fixed an eyepiece for direct observations. A telescope using this arrangement is known as a **Newtonian telescope**.

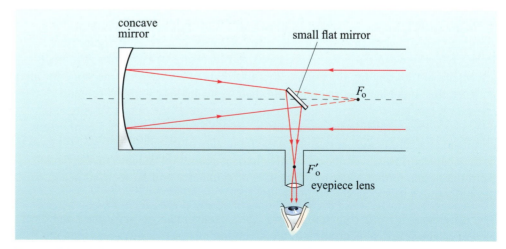

Figure 2.8   A Newtonian (reflecting) telescope. Parallel rays of light would be brought to a focus at F_o, but are intercepted by a small flat mirror. This moves the focal point to one side, at F_o', before the rays are rendered parallel by the eyepiece lens. The final virtual image is located at infinity and is inverted with respect to the object.

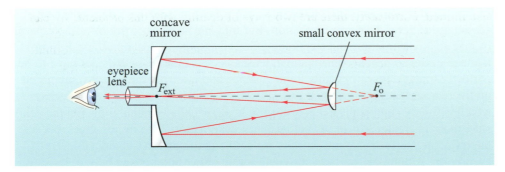

(a)

Figure 2.9 A Cassegrain (reflecting) telescope. Parallel rays of light would be brought to a focus at F_0, but are intercepted by a small convex mirror. This makes the rays of light diverge somewhat so that they are not brought to a focus until the point F_{ext}. The rays then diverge before entering the eyepiece lens and emerge from it parallel. The final virtual image is once again at infinity and inverted.

A further improvement was introduced by the French astronomer Guillaume Cassegrain, one of Newton's contemporaries. His idea is illustrated in Figure 2.9 and is now used in many large modern telescopes. In place of Newton's flat and tilted secondary mirror, Cassegrain used a slightly *diverging* secondary mirror placed on the optical axis of the primary mirror. The light is therefore reflected back towards the centre of the primary mirror, where it passes through a hole on the optical axis and then onto an eyepiece. This has the effect of extending the path of the reflected light before it is brought to a focus at F_{ext}. The **effective focal length** of the system of two mirrors is the focal length of a single mirror having the same diameter as the objective and giving a cone of light converging at the focus at the same angle as the two-mirror system. It is the effective focal length of the optical system which determines the size of the image, and in a **Cassegrain telescope** the effective focal length can be many times that of a Newtonian telescope of the same length. Both Newtonian and Cassegrain telescopes may be constructed using either paraboloidal objective mirrors or using spherical objective mirrors with Schmidt correcting plates.

If a telescope is to be used with a photographic or electronic detector (see Chapter 4), instead of the eye, then we must allow a real image to fall onto the light-sensitive surface of the detector. In this case there is no point in using the telescope with an eyepiece, since that produces a virtual image located at infinity. (Remember, when you are actually *looking* through a telescope, the very final image is that produced by the lens of your eye. This image falls onto your retina and is therefore a real image. However, the final image produced by the *telescope*, with the eyepiece in place, is a virtual image, located at infinity.) The simplest solution is to remove the eyepiece entirely and place the detector in the focal plane of the mirror system (i.e. at F'_0 in Figure 2.8 or at F_{ext} in Figure 2.9). This also has the advantage of removing any aberrations introduced by the eyepiece lens. Alternatively, the secondary mirror may also be removed, and the detector may be placed directly at the **prime focus** of the main mirror (i.e. at F_0 in Figure 2.8 or at F_0 in Figure 2.9). This has the additional advantage of removing one more optical component, and with it the inherent aberrations and absorption losses that it contributes. Figure 2.10a shows a photograph of a modern optical (reflecting) telescope made to a Cassegrain design with an 8 m diameter parabolic primary mirror. Figure 2.10b shows a much smaller Schmidt–Cassegrain telescope, of a type you may use as a student.

(b)

Figure 2.10 (a) The 8 m diameter Gemini (North) Telescope on the island of Hawaii. This is a Cassegrain telescope with a parabolic primary mirror. (*Photo courtesy of Gemini Observatory*) (b) A 40 cm diameter Schmidt–Cassegrain telescope, of the type often used for student projects.

In comparison with refracting telescopes, the reflectors start with the important advantage of zero chromatic aberration. But they also score heavily on some aspects of practical construction and technology. For very large diameters (10 m or more) it is much easier to produce mirrors than lenses because the glass does not have to be perfectly transparent or optically homogeneous and a mirror can be fully supported on the rear surface. The grinding and polishing is carried out on only one surface, which is finally covered by a thin reflecting layer, usually of aluminium. On the debit side, there is greater loss of optical intensity in reflectors than in refractors, because the reflecting surfaces are never 100% reflective and may have appreciable absorption. Aluminized surfaces also deteriorate rather quickly and have to be renewed every few years. On the other hand, a perfectly polished lens remains serviceable for many years.

2.4 The characteristics of astronomical telescopes

Having looked at the different designs of optical telescopes and the various problems inherent in their construction, we now turn to the ways in which their performance may be characterized. We consider five main performance characteristics, each of which may be applied to both refracting telescopes and reflecting telescopes.

2.4.1 Light-gathering power

One of the key benefits of using a telescope is that it enables fainter objects to be detected than with the naked eye alone. The **light-gathering power** of a simple telescope used with an eyepiece is defined as

$$\text{light-gathering power} = (D_o/D_p)^2 \qquad (2.1)$$

where D_o is the diameter of the objective lens (or mirror) and D_p is the diameter of the eye's pupil, assuming that all the light passing through the objective enters the eye. This is proportional to the light-gathering area of the objective lens or mirror of the telescope.

■ Compare the light-gathering powers of three telescopes with objective mirrors of diameter $D_o = 5$ cm, 25 cm and 1 m. Assume that the eye has a pupil diameter of $D_p = 5$ mm.

❑ The light-gathering power of a telescope is given by the ratio $(D_o/D_p)^2$. Hence for the three telescopes we have, (converting all diameters to mm):

For $D_o = 5$ cm $\qquad (D_o/D_p)^2 = (50 \text{ mm}/5 \text{ mm})^2 = 10^2$

For $D_o = 25$ cm $\qquad (D_o/D_p)^2 = (250 \text{ mm}/5 \text{ mm})^2 = 2.5 \times 10^3$

For $D_o = 1$ m $\qquad (D_o/D_p)^2 = (1000 \text{ mm}/5 \text{ mm})^2 = 4 \times 10^4$.

Clearly, the larger the aperture the more light is collected and focused into the image, and therefore fainter stars can be detected.

2.4.2 Field-of-view

The **field-of-view** of a telescope is the angular area of sky that is visible through an eyepiece or can be recorded on a detector, expressed in terms of an angular diameter. When a telescope is used with an eyepiece, the angular field-of-view is equal to the diameter of the **field stop** (i.e. the diameter of the aperture built into the eyepiece) divided by the effective focal length of the primary mirror or lens. In symbols:

$$\theta = D/f_o \qquad (2.2)$$

where the angular diameter of the field-of-view, θ, is in radians.

■ What is the field-of-view, in arcminutes, of a telescope whose focal length is 3050 mm when used with an eyepiece with a field-stop diameter of 23.0 mm?

❏ The angular diameter of the field-of-view is $\theta = D/f_o = 23.0 \text{ mm}/3050 \text{ mm} = 7.54 \times 10^{-3}$ radians. Converting to degrees, this is 7.54×10^{-3} radians $\times (180/\pi)$ degrees radian$^{-1} = 0.432° = 25.9'$.

Remember that π radians corresponds to 180°.

When a telescope is used with a detector in place of an eyepiece, the determining factor here is the linear size of the detector itself, rather than the field-stop diameter.

■ What is the maximum focal length telescope that could accommodate a 1° field-of-view on a standard 35 mm film frame?

❏ $1° = (\pi/180)$ radians, so the limiting focal length is given by 35 mm/$(\pi/180) = 2005$ mm, or about 2.0 m. A focal length longer than this would reduce the field-of-view.

2.4.3 Angular magnification

You may be familiar with the scales that appear on terrestrial maps or images obtained with microscopes, possibly stated as 1 : 100 000 or 1 mm corresponds to 1 μm. Scales such as these indicate how the size of the reproduction compares to the real thing. Image scales are no less important in astronomy, though they are usually stated in a different form, as we now explain. Imagine for a moment that you have the use of a telescope that allows you to observe Saturn and its ring system. It must be very highly magnified to show so much detail, mustn't it? Well, consider the size of the image. It is in fact greatly *de*magnified, by such a large factor that the image of the 120 000 km diameter planet fits on the light-sensitive surface of your eye only a few millimetres across. The same would be true if you recorded the image on photographic film or with a digital camera. Yet you know you can see more detail than with the naked eye. This simple example emphasizes that the important magnification in much astronomical imaging is not the *linear* magnification described above for terrestrial maps, but rather the *angular* magnification. The angular magnification indicates by what factor the angular dimension (e.g. angular diameter) of a body is increased. So if you were to observe Saturn through a telescope, you would be benefiting from a high angular magnification which makes the image appear larger even though it is squeezed into the tiny space of your eyeball.

The **angular magnification** M of an astronomical telescope, used visually, is defined as the angle subtended by the image of an object seen through a telescope, divided by the angle subtended by the same object without the aid of a telescope. By geometry, this can be shown to be equivalent to

$$M = f_o/f_e \qquad (2.3)$$

where f_o is the effective focal length of the objective lens or mirror system and f_e is the focal length of the eyepiece lens.

■ What is the angular magnification of a Newtonian reflecting telescope with a mirror of focal length $f_o = 10$ m and an eyepiece of focal length $f_e = 10$ cm?

❏ The angular magnification is equal to the ratio f_o/f_e. Thus we have $M = (10 \text{ m})/(0.1 \text{ m}) = 100$. The larger the focal length of the primary mirror, the greater will be the angular magnification of the telescope.

Notice that the angular magnification and field-of-view of a telescope both depend on the focal length of the objective lens or mirror. However, increasing f_o will *increase* the angular magnification but *decrease* the field-of-view, and vice versa.

2.4.4 Image scale

The nearest equivalent definition to angular magnification that is applicable to telescopes used for imaging onto a detector is the **image scale** (sometimes called the *plate scale*). Because of the importance of angular measures, the image scale quoted by astronomers indicates how a given angular measure on the sky corresponds to a given physical dimension in an image. The most common convention is to state how many arcseconds on the sky corresponds to 1 mm in the image.

Fortunately, it is very easy to calculate the image scale for any imaging system, as it depends on only one quantity: the focal length f_o of the imaging system. The image scale I in arcseconds per millimetre is given by

$$I \text{ / arcsec mm}^{-1} = \frac{1}{(f_o \text{ / mm}) \times \tan(1 \text{ arcsec})} \tag{2.4}$$

Note that as the image on the detector becomes *larger*, the numerical value of I becomes *smaller*.

■ A certain telescope has an objective with an effective focal length of 3000.0 mm. What is the image scale in the image plane?

❏ First note that $\tan(1 \text{ arcsec}) = 4.848 \times 10^{-6}$, so $1/\tan(1 \text{ arcsec}) = 206\,265$. Hence, $I = (206\,265/3000.0) \text{ arcsec mm}^{-1} = 68.755 \text{ arcsec mm}^{-1}$. (For small angles, $\tan\theta \approx \theta$ in radians.)

The number 206 265 is quite useful in astronomy, as it is the number of arcseconds in 1 radian of angular measure, given by $(180 \times 60 \times 60)/\pi$.

2.4.5 Point spread function and angular resolution

The image of a point-like source of light (such as a distant star) obtained using a telescope will never be a purely point-like image. Even in the absence of aberrations and atmospheric turbulence to distort the image, the image of a point-like object will be extended due to diffraction of light by the telescope aperture. The bigger the aperture, the smaller is the effect, but it is still present nonetheless. The intensity of the image of a point-like object will take the form shown in Figure 2.11a. The structure shown here is referred to as the **point spread function** (PSF) of the telescope. Lens or mirror aberrations and atmospheric turbulence will each cause the width of the PSF to broaden, and may cause its shape to become distorted too. However, in the ideal case when neither aberrations nor turbulence is present, the telescope is said to be **diffraction-limited**, and its PSF has the form shown. The

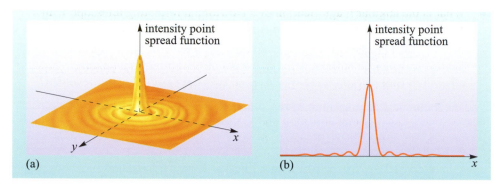

Figure 2.11 The image of a point-like object is not point-like even under ideal conditions. (a) The vertical direction represents image intensity. The point spread function of a point-like object under ideal conditions consists of a central peak surrounded by concentric ripples. The two-dimensional PSF has the circular symmetry of the telescope aperture. (b) A slice through (a) along one axis.

width of the PSF, in this idealized case, is inversely proportional to the aperture diameter of the telescope.

Using the idea of the diffraction-limited PSF, we can also define the (theoretical) **limit of angular resolution** for an astronomical telescope. This is the minimum angular separation at which two equally bright stars would just be distinguished by an astronomical telescope of aperture D_o (assuming aberration-free lenses or mirrors and perfect viewing conditions). As shown in Figure 2.12b, at a certain separation, the first minimum of the PSF of one star will fall on the peak of the PSF of the other star. At this separation, the two stars are conventionally regarded as being just resolved.

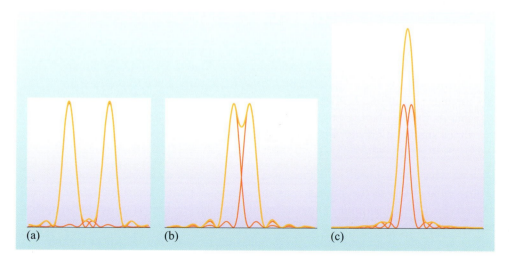

Figure 2.12 The images of the two stars in (a) are clearly resolved, whereas those in (c) are unresolved. In (b), the first minimum of one PSF coincides with the peak of the other PSF. At this separation the stars are said to be just resolved.

The angular separation corresponding to the situation in Figure 2.12b is given by

$$\alpha_c = 1.22\lambda/D_o \qquad (2.5)$$

where α_c is the limit of angular resolution measured *in radians* and λ is the average wavelength of light contributing to the image. As noted above, the limit of angular resolution arises due to diffraction of light by the telescope aperture and represents a fundamental limit beyond which it is impossible to improve.

Equation 2.5 is often known as the Rayleigh criterion.

■ A certain ground-based reflecting telescope contains aberration-free optical components and has a primary mirror aperture 500 mm in diameter. If two stars of equal brightness are observed through a red filter that transmits only light of wavelength 650 nm, what is the *theoretical* minimum angular separation (in arcseconds) at which these two stars could be just resolved?

❏ The theoretical angular limit of resolution is given by the formula

$$\alpha_c = 1.22\lambda/D_o,$$

hence in this case:

$$\alpha_c = (1.22 \times 650 \times 10^{-9} \text{ m})/(0.50 \text{ m}) = 1.6 \times 10^{-6} \text{ radians}$$

So, $\alpha_c = (1.6 \times 10^{-6} \times 180/\pi)$ degrees $= 9.1 \times 10^{-5}$ degrees.

Or, $\alpha_c = 9.1 \times 10^{-5} \times 3600$ arcseconds $= 0.33''$.

In practice, two stars this close together are unlikely to be resolved using a conventional ground-based telescope, whatever its aperture diameter, because of the degradation of angular resolution imposed by turbulence in the atmosphere. This makes all single stars appear to be of a small, but finite, angular size, typically of order 1″ across. In effect therefore, atmospheric turbulence *broadens* the PSF of the telescope. In fact, in most ground-based observatory telescopes, the dominant contribution to the size of the PSF is generally from atmospheric turbulence rather than imperfections in the telescope optics or the theoretical limit to angular resolution imposed by diffraction. Hence, the diameter of the actual point spread function is a common way of quantifying the astronomical **seeing**. At the very best, the seeing from a good astronomical site is around 0.5″, but at most observing sites it may be a few arcseconds even on good nights.

■ If it's never possible to achieve a diffraction-limited point spread function, because of atmospheric turbulence, what's the point of building a ground-based optical telescope with a mirror diameter of 5 m or more?

❏ A 5 m mirror will have a theoretical limit of angular resolution of about 0.03″ (i.e. 10 times smaller than the example above, due to its 10 times larger mirror). This will be degraded by atmospheric turbulence to produce an angular resolution of order ~1″. However, the advantage of the 5 m mirror is that its light-gathering power is 100 times greater than that of a mirror of 10 times smaller diameter. So much fainter astronomical objects may be detected.

Despite what has just been said, there is a technique now available at some professional observatories for reducing the effects of poor seeing, and attaining close to the theoretical limit of angular resolution. The technique of **adaptive optics** refers to a process whereby corrections to the shape of the primary or secondary mirror are made on a rapid timescale (hundredths of a second) to adjust for the image distortions that arise due to atmospheric turbulence. A relatively bright reference star is included within the field-of-view, or an artificial *laser guide star* is produced by directing a laser into the atmosphere. The adaptive optics system then rapidly adjusts the mirror under software control in order to make the size of the PSF of the reference star as small as possible. By correcting the reference star in this way, all other objects in the field-of-view have their PSFs similarly corrected, and an angular resolution close to the theoretical limit may be obtained.

2.5 Telescope mountings

A telescope with the largest light-gathering power, best point spread function and optimum image scale and field-of-view is of little use unless it is mounted in an appropriate way for tracking astronomical objects across the night sky. It is essential that a telescope can be pointed accurately at a particular position in the sky and made to track a given position as the Earth rotates on its axis as noted in Chapter 1.

Broadly speaking there are two main types of mounting for astronomical telescopes, known as alt-azimuth and equatorial. An **alt-azimuth mounting** (alt-az for short) is the simplest to construct. It allows motion of the telescope in two directions, namely the altitude or vertical direction and the azimuth or horizontal direction (Figure 2.13a). Although simple and relatively cheap to construct, it has the drawback that to accurately track an astronomical object such as a star or galaxy requires the telescope to be driven in *both* axes simultaneously at varying speeds. Given the widespread availability of computer software to do the job this is not a problem in practice. However, another limitation is that the image will rotate as the telescope tracks, and therefore the detector must also be counter-rotated during any exposure in order to produce an un-trailed image.

The other type of mounting is known as an **equatorial mounting**. In this case, one axis of the telescope is aligned parallel to the rotation axis of the Earth (the so called **polar axis**), and the other axis (the so called **declination axis**) is at right angles to this (Figure 2.13b). This has the advantage that once the telescope is pointed at a particular star or galaxy, then tracking of the object as the Earth rotates is achieved simply by moving the telescope at a constant speed around the polar axis only.

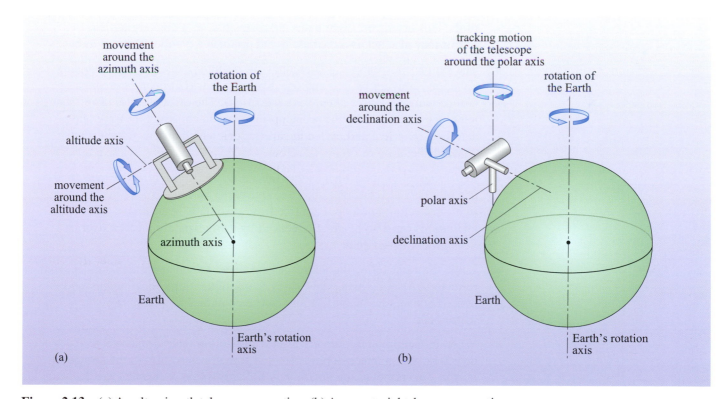

Figure 2.13 (a) An alt-azimuth telescope mounting. (b) An equatorial telescope mounting.

■ For a given location on the Earth, what determines the angle between the polar axis of an equatorially mounted telescope and the horizontal?

❏ The angle between the polar axis and the horizontal is equal to the latitude of the location.

An equatorial mounting is relatively expensive to construct, but it is much simpler to drive and point a telescope with such a mounting, particularly without computer assistance, and the field does not rotate during the course of an exposure.

■ At what speed must a telescope be moved around the polar axis of an equatorial mounting in order to counteract the effect of the Earth's rotation?

❏ The telescope must be rotated at a rate of one revolution (360°) for every sidereal day (23 hours 56 minutes 4 seconds) in the opposite sense to that in which the Earth rotates.

2.6 Summary of Chapter 2 and Questions

- *Converging* lenses or mirrors cause parallel beams of light to be brought to a focus at the *focal point*, situated at a distance of one *focal length* beyond the lens or one focal length in front of the mirror. *Diverging* lenses or mirrors cause parallel beams of light to diverge as if emanating from the focal point of the lens or mirror. Light paths are reversible, so a converging lens or mirror may also act as a *collimator* and produce a parallel beam of light.

- The simplest astronomical telescopes are *refracting* telescopes comprising either one converging lens and one diverging lens (*Galilean telescope*), or two converging lenses (*Keplerian telescope*). The effectiveness of refracting telescopes is limited by the problems involved in constructing large lenses, and their *spherical* and *chromatic* aberrations which are, to some extent, unavoidable.

- *Reflecting* telescopes, such as the *Newtonian* and *Cassegrain* designs, make use of a curved (concave) objective (primary) mirror to focus the incoming light. Reflecting telescopes are free from chromatic aberrations. Spherical aberrations can also be greatly reduced by using a paraboloidal mirror or a *Schmidt* correcting plate.

- Large-diameter reflecting telescopes are easier to construct than similar sized refractors. Also, by using the Cassegrain design, a long focal length (and hence high angular magnification) can be contained in a relatively short instrument.

- When reflecting telescopes are used with photographic or electronic detectors, the eyepiece is removed, and sometimes so also is the secondary mirror. This removes the aberrations and absorption losses that are due to these components and allows a real image to fall directly onto the light-sensitive surface of the detector.

- The main parameters of an optical telescope are its light-gathering power, its field-of-view, its angular magnification or image scale and its limit of angular resolution.

- The angular size of the *point spread function* of a telescope can be used to quantify the astronomical *seeing*. The technique of *adaptive optics* can compensate for the effects of atmospheric turbulence and produce images whose PSFs are close to being *diffraction-limited*.

- A telescope may have an *alt-azimuth* or *equatorial* mounting. The former is less complex to construct, but with the latter it is simpler to point and drive a telescope.

QUESTION 2.1

Summarize how the following characteristics of a visual telescope

(i) light-gathering power,

(ii) field-of-view,

(iii) angular magnification,

(iv) limit of angular resolution,

depend on the aperture D_o and the focal length f_o of its objective lens (for a given eyepiece of focal length f_e).

QUESTION 2.2

(a) Calculate the ratio of the light-gathering power of a reflecting telescope of diameter $D_o = 5.0$ m to that of a refractor of diameter 1.0 m (neglect losses of light, mentioned in the text).

(b) Compare the (theoretical) limits of angular resolution of these two telescopes (at the same wavelength).

QUESTION 2.3

(a) The atmospheric seeing at a particular observatory site is 1 arcsecond (1″). What is the aperture of a *diffraction-limited* telescope (at a wavelength of 485 nm) which would have a resolving power equivalent to this seeing?

(b) Why then do you think that astronomers build such large and expensive telescopes for use in ground-based observations?

QUESTION 2.4

List the important advantages and disadvantages of reflecting telescopes compared to refracting telescopes.

QUESTION 2.5

What is the Schmidt correcting plate and how does it improve the performance of a reflecting telescope? (b) Draw a diagram illustrating how a Cassegrain telescope equipped with a Schmidt correcting plate focuses light from a distant object.

QUESTION 2.6

(a) Calculate the image scale in the focal plane of a 300 mm diameter telescope whose optical system is stated as F/10.

(b) The angular diameter of the planet Mars varies from about 14″ to 25″ depending on its distance from the Earth. Calculate how large the image would be in the focal plane of a 300 mm diameter, F/10 telescope at its closest and furthest.

3 SPECTROGRAPHS

Telescopes may simply be used to collect the light from an astronomical object in order to measure its position, brightness or spatial distribution. However, it is often far more instructive to examine the **spectrum** of light from an object such as a star or galaxy, namely the distribution of light intensity as a function of wavelength.

3.1 The heart of spectroscopy: dispersing light

The spectrum of a light source may be revealed in several ways, all of which involve making light of different wavelengths travel in different directions, a process which we term **dispersion**. There are two principal ways of dispersing light: using either a *prism* or a *grating*.

3.1.1 Prisms and the refraction of light

The simplest way to disperse light is to use a **prism**. When light enters a prism, it is no longer travelling in a vacuum, and its speed decreases. If the incident wavefront is travelling at an angle to the surface of the prism, which is easy to arrange because of its angled faces, then the propagation of the part of the wavefront in the prism is retarded, thus bending the wavefront and changing its direction of propagation through the prism (Figure 3.1). This phenomenon is referred to as **refraction**.

The speed of light in most materials depends on frequency, so the change in direction also depends on frequency, and hence different colours become separated. Figure 3.2 illustrates the situation when a beam of white light (i.e. a mixture of all colours) encounters a triangular glass prism. The white light is dispersed at the air–glass boundary and, because of the shape of the prism, the different colours undergo *further* dispersion at the glass–air boundary as they leave the prism.

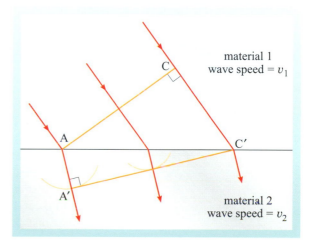

Figure 3.1   The wavefront AC is incident upon the surface AC′. In the time that it takes the wavefront to travel the distance CC′ in material 1, the wavefront has travelled a shorter distance AA′ in material 2, thus changing the direction of propagation. The distance AA′ depends on the speed of light in material 2, which depends on the frequency of the radiation, and hence the amount of refraction also depends on frequency.

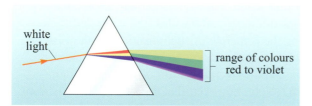

Figure 3.2   A beam of white light enters a triangular prism as shown. Red light has a lower frequency (longer wavelength) than violet light and the direction of the red beam is altered less than the direction of the violet beam. Consequently the white light spreads into its constituent colours within the prism. The different colours are further dispersed on leaving the prism.

3.1.2 Diffraction and interference of light

When light, or indeed any type of wave, passes through a narrow aperture, it will spread out on the other side. This is the phenomenon of **diffraction**. For example Figure 3.3 shows the diffraction of water waves in a device called a ripple tank. The extent to which waves are diffracted depends on the size of the aperture relative to the wavelength of the waves. If the aperture is very large compared to the wavelength, then the diffraction effect is rather insignificant. So although sound waves may be diffracted by a doorway, light waves are not appreciably diffracted by doorways because the wavelength of visible light (about 400 to 700 nm) is very small in comparison to the width of the doorway. But light *is* diffracted, and provided the slit is narrow enough, the diffraction will become apparent.

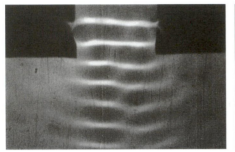 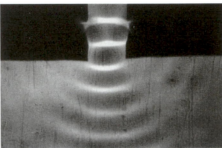 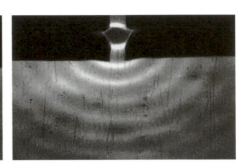

Figure 3.3 The diffraction of water waves in three different cases. As the slit width is reduced, the amount of diffraction increases.

The phenomenon of diffraction allows us to appreciate the effect of an aperture on the propagation of waves, however it says nothing about what will happen when waves from different sources or from different parts of the same source meet. For this, the **principle of superposition** must be used. The principle of superposition states that if two or more waves meet at a point in space, then the net disturbance at that point is given by the sum of the disturbances created by each of the waves individually. For electromagnetic radiation the disturbance in question can be thought of as variations in electric and magnetic fields. The effect of the superposition of two or more waves is called **interference**.

To begin with, we consider the diffraction of monochromatic light by a pair of closely spaced, narrow slits as shown in Figure 3.4. Plane waves of constant wavelength from a single, distant, source are diffracted at each of two slits, S_1 and S_2. Because the waves are from the same original source they are in phase with each other at the slits. At any position beyond the slits, the waves diffracted by S_1 and S_2 can be combined using the principle of superposition. In the case of light waves, the resulting illumination takes the form of a series of light and dark regions called **interference fringes** and the overall pattern of fringes is often referred to as a **diffraction pattern**. (Note, however, that the same pattern is also sometimes referred to as an *interference pattern*. The reason for the dual nomenclature is that both diffraction *and* interference are necessary in order to generate the observed pattern, so either is an appropriate description.)

In order to appreciate how the interference fringes arise consider Figure 3.4a. When the wave arriving at a point on the screen from slit S_1 is in phase with the wave arriving from S_2, the resultant disturbance will be the sum of the disturbances caused by the waves individually and will therefore have a large amplitude (as shown in Figure 3.5a). This is known as **constructive interference**. When the waves are completely out of phase, the two disturbances will cancel. This is known as **destructive interference** (as shown in Figure 3.5b).

On the screen, constructive interference will cause relatively high intensity, while destructive interference will lead to low intensity, hence the observed pattern of fringes.

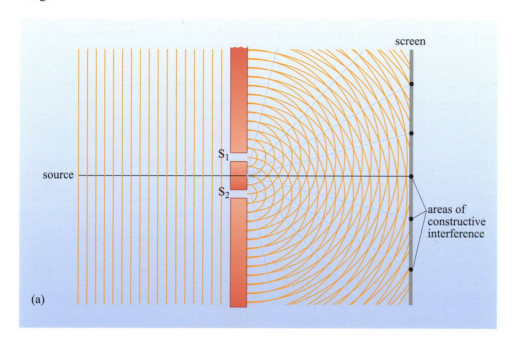

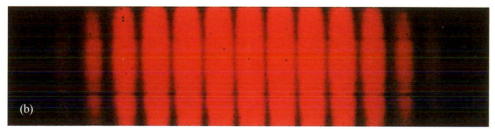

Figure 3.4 (a) Diffraction and interference of light produced by two narrow slits S_1 and S_2. (b) Bright and dark interference fringes on the screen.

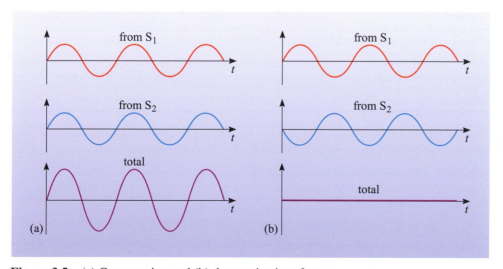

Figure 3.5 (a) Constructive and (b) destructive interference.

The general condition for constructive interference at any point is that the **path difference** between the two waves is a whole number of wavelengths, i.e.

$$\text{path difference} = n\lambda, \text{ where } n = 1, 2, 3\ldots \quad (3.1)$$

The general condition for destructive interference is that the path difference is an odd number of half-wavelengths, i.e.

$$\text{path difference} = (n + \tfrac{1}{2})\lambda, \text{ where } n = 1, 2, 3\ldots \quad (3.2)$$

The result of this is that when a source of light consisting of a *range* of wavelengths is used, the positions of constructive interference will be *different* for each wavelength. In other words, the combination of diffraction and interference produced by a pair of slits has the effect of dispersing light into its constituent wavelengths.

The same principles also apply when not two but a large number of equally spaced slits are used. Such **diffraction gratings** typically have several hundred slits per millimetre and give much sharper diffraction patterns than a simple double slit. The individual fringes can also be much further apart, so that dispersed wavelengths can be more widely separated. The details of this phenomenon, as applied to astronomical spectroscopy, are discussed below.

3.1.3 Reflective diffraction gratings

Although the above description of diffraction has been in terms of light passing through a series of slits in a (transmission) diffraction grating, the type of grating which is currently most common in astronomy is a **reflective diffraction grating** or **reflection grating**. This again exploits the wave properties of light, in this case by making adjacent sections of a wavefront travel extra distances as it is reflected off a non-uniform surface. The non-uniform surface is actually a very precisely made mirror into which steps or grooves have been cut; as shown by the cross-section in Figure 3.6. The wavefront propagating from groove A and the wavefront propagating from groove B will constructively interfere with each other only if the difference in the lengths of the light paths, from L to L″, is an integer number of wavelengths. From the figure, the path difference is $d \sin \alpha + d \sin \beta$, and the condition for constructive interference can therefore be written

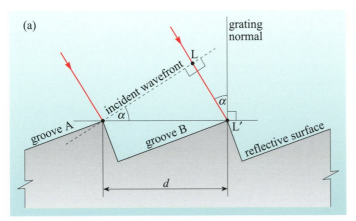
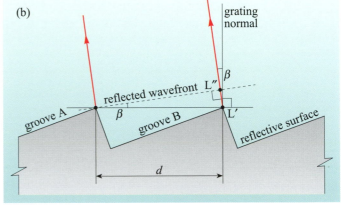

Figure 3.6 (a) For a wavefront incident on the grating at an angle α to the normal, the portion of the wavefront reaching groove B has to travel an extra distance LL′ = $d \sin \alpha$ compared to the portion of the wavefront reaching groove A, where d is the distance between successive grooves. (b) Similarly, for light reflected from the grating at an angle β to the normal, the portion of the wavefront reflected from groove B has to travel an extra distance L′L″ = $d \sin \beta$. The total path difference from L to L″ is therefore $d(\sin \alpha + \sin \beta)$.

$$d(\sin \alpha + \sin \beta) = n\lambda \qquad (3.3)$$

This is such an important equation for astronomers that it is given a name, the **grating equation**. The integer n is called the **spectral order**, and quantifies how many wavelengths of path difference are introduced between successive grooves on the grating.

Now consider the grating equation. The groove spacing d is a feature of the grating, and the angle of the incident light α will be the same for all wavelengths, so the only remaining variables are the diffraction angle β and the wavelength λ. It is therefore clear that β must depend on wavelength, which is to say that the grating is a means of sending light of different wavelengths in different directions, i.e. producing a spectrum.

■ Imagine you have a grating spectrograph whose grating has 1000 grooves per mm, and is set up with the light incident at an angle of 15° to the grating normal. Calculate the angles at which light of: (i) 400 nm, (ii) 500 nm, and (iii) 600 nm will be diffracted in the first spectral order. You may find it convenient to express the wavelength and the groove spacing d in units of microns.

❏ Rearrange the grating equation $d(\sin \alpha + \sin \beta) = n\lambda$ and write

$\sin \beta = n\lambda/d - \sin \alpha$

Then substitute in the values $d = 0.001$ mm $= 1\ \mu$m, $\alpha = 15°$ and $n = 1$ to give $\sin \beta = \lambda/\mu$m $- 0.2588$. We can then calculate the diffraction angles β as follows:

For $\lambda = 400$ nm $= 0.4\ \mu$m, we have $\sin \beta = 0.1412$, therefore $\beta = 8.1°$.

For $\lambda = 500$ nm $= 0.5\ \mu$m, we have $\sin \beta = 0.2412$, therefore $\beta = 14.0°$.

For $\lambda = 600$ nm $= 0.6\ \mu$m, we have $\sin \beta = 0.3412$, therefore $\beta = 20.0°$.

One feature of the spectrum produced by a diffraction grating is that multiple spectra are produced, corresponding to different spectral orders. For example, it is obvious from the grating equation that for a given spectrograph set-up, i.e. for some particular values of d and α, light at 700 nm in the first spectral order ($n = 1$) travels at the same diffraction angle β as light at 350 nm in the second spectral order ($n = 2$). Depending on the sensitivity of the detector and the relative flux in the source at overlapping wavelengths, it may be necessary to use a filter to block out the unwanted wavelengths.

It is instructive to ask how the choice of grating and spectral order affects the dispersion of the spectrum, i.e. the amount by which the light is spread out. The **angular dispersion** is a measure of how large a change $\Delta\beta$ in the diffraction angle results from a change $\Delta\lambda$ in wavelength, so we want to know $\Delta\beta/\Delta\lambda$. Calculus makes the calculation of $\Delta\beta/\Delta\lambda$ very straightforward, so we shall use that approach here. The grating equation can be rearranged as $\sin \beta = n\lambda/d - \sin \alpha$. Using calculus, we can then write

If your calculus is too rusty for you to follow this, then skip the steps and just note the result.

$$\frac{\partial \beta}{\partial \lambda} = \frac{\partial \beta}{\partial \sin \beta}\frac{\partial \sin \beta}{\partial \lambda} = \frac{1}{\partial \sin \beta / \partial \beta}\frac{\partial \sin \beta}{\partial \lambda} = \frac{1}{\cos \beta}\frac{n}{d} \qquad (3.4)$$

This indicates that the angular dispersion can be increased by working in higher spectral orders, i.e. by increasing n, or by using gratings with narrower groove spacings d (i.e. more grooves per millimetre such that d is smaller). Hence a grating with 600 grooves per millimetre will have twice the dispersion of a grating with 300 grooves per millimetre, if they are both used in the same spectral order. Of course, a grating of 300 grooves per millimetre used in second order ($n = 2$) will give the same dispersion as a grating with 600 grooves per millimetre used in first order ($n = 1$).

In the last few years, a new type of diffraction grating has become common in astronomy. The reflective diffraction grating described above works by introducing a different path length between parts of a wavefront striking different grooves of the grating. A **volume phase holographic grating** (VPH grating), in contrast, is a transparent medium, usually a layer of gelatine sandwiched between two glass plates. The refractive index of the gelatine varies in a carefully defined way from point to point. VPH gratings can offer superior efficiency and versatility to reflection gratings, and can be produced in the much larger sizes needed for the next generation of large telescopes.

3.2 Optical layout of a spectrograph

Although spectrograph designs vary widely, most consist of a few key elements which we describe below. You may be surprised to learn that a slit is *not* a key element of a spectrograph, but it is often a useful one, as we shall see later.

We begin in the focal plane of the telescope, where an image of the sky is formed by the telescope optics. The rays of light associated with each object are converging as they approach the focal plane, and they diverge beyond it (Figure 3.7). In this condition they are not suitable for dispersing; first they must be rendered parallel. This is achieved with a **collimator**, which is a lens or mirror of focal length f_{col} placed a distance f_{col} beyond the focal plane. The focal plane of the telescope is therefore at the focal point of the collimator, and hence the rays of light emerging from the collimator are made parallel to one another, i.e. **collimated**.

The collimated beam is then dispersed using one of the dispersing elements discussed previously, i.e. a prism, reflective diffraction grating, or VPH grating. The dispersing element takes the bundle of parallel rays, and produces a separate parallel bundle for each wavelength present in the object, each bundle travelling in a slightly different direction. Each bundle consists of light of a single wavelength, i.e. it is **monochromatic**.

The diverging bundles of parallel rays must now be brought to a focus. This is achieved with another lens or mirror called a **camera** lens or mirror, whose focal length is f_{cam}, and which therefore focuses each bundle of parallel rays onto its image plane a distance f_{cam} beyond the lens or mirror. Because each monochromatic bundle is travelling in a different direction, each is focused onto a different part of the image plane. The image in this plane is therefore a spectrum.

Note that we have not discussed a spectrograph slit at all, and that is intentional. From the discussion above, you should see that a spectrograph produces a new image of whatever image lies in the telescope focal plane, but in a new position that depends on its wavelength. If you had a star-like object giving out all its light at only *three wavelengths*, then in the spectral image plane you would see three round star-like images, each corresponding to one of the three different wavelengths

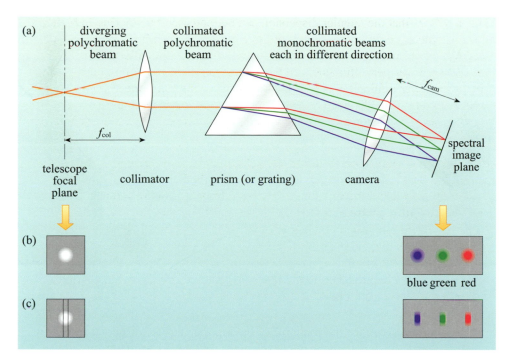

Figure 3.7 (a) Spectrograph layout showing the focal plane of the telescope, the collimator lens, dispersive element (i.e. prism or grating), camera lens, and spectral image plane. (b) Views of a 'star' emitting three strong emission lines, first in the focal plane of the telescope where a single polychromatic image can be seen, and secondly in the spectral image plane where three separated monochromatic images are visible. (c) As for (b), but with a slit placed in the focal plane of the telescope to restrict the width of each monochromatic image of the star in the spectral image plane.

(see Figure 3.7b). If these three wavelengths are very similar to one another, then the three images in the spectrum would not be very far apart. Indeed they might be so close together that they were almost indistinguishable.

How then could you tell how much light there is at one wavelength compared with another? You would have two choices. You might be able to increase the dispersion to move the three images further apart, either by observing at a higher spectral order or by using a grating with a finer groove spacing, as was discussed earlier. If this is not practical, then you have only one other possibility, which is to block off part of each monochromatic image so that they no longer overlap. This is achieved by putting a mask in the focal plane of the telescope where the polychromatic (white-light) image is first formed. The mask needs to block off only the edges of the image, not the top and bottom as well, so a long, narrow mask can be used. Astronomers call such a mask a **slit** or entrance slit. The result of doing this is shown in Figure 3.7c.

The spectra of real astronomical objects do not just consist of three wavelengths, but often contain a number of spectral lines (either in absorption or emission), superimposed on a continuum spectrum. In order to distinguish between spectral lines of similar wavelength, a suitably narrow slit is needed. If the slit is made narrower, each monochromatic image, and hence each spectral line, becomes narrower and hence more easily distinguished. Unfortunately, the amount of light allowed to enter the spectrograph is also reduced, and hence the intensity of the image decreases.

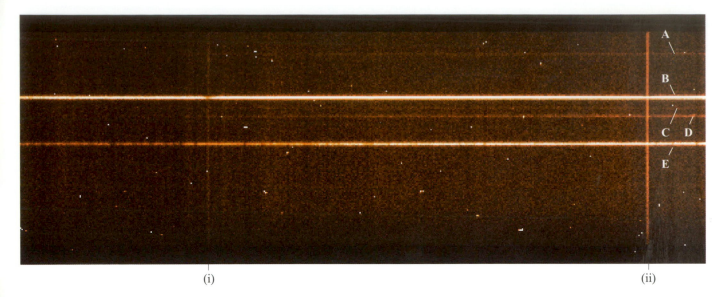

(i) (ii)

Figure 3.8 An image showing the spectra of two bright stars (B and E) and three faint ones (A, C and D). The spectral direction runs horizontally, and the spatial direction (along the slit) runs vertically in this image. The two bright vertical lines, (i) and (ii), correspond to two emission lines in the spectrum of the sky, which fills the slit.

Astronomers often face the trade-off of having the slit narrow enough to allow them to distinguish neighbouring wavelengths, but wide enough to provide an adequate number of photons. In practice, the width of the slit is often matched to the seeing conditions.

The result of using a slit like this is shown in Figure 3.8, where the spectra of several stars that lie along the slit in the focal plane of the telescope are all recorded.

3.3 Throughput of a spectrograph

Modest sized telescopes (with aperture diameters of up to a few tens of cm) allow the spectra of bright stars to be obtained. Although faint stars can be seen through the telescope, once their light is dispersed by the diffraction grating or prism, the amount of light falling on any single part of the image plane is greatly reduced. This is because the light is spread out in the spectrum, though in principle this does not lead to any loss of photons. However, the large number of optical elements in the spectrograph does contribute to real losses of intensity. For example, the entrance slit sometimes has to be set narrower than the apparent size of the star due to atmospheric seeing, with the result that only a fraction of the starlight enters the spectrograph.

- ■ Why are you not completely free to make the entrance slit wider to allow more light through?

- ❏ Making the slit wider also makes the spectral lines wider in the imaged spectrum, so using a wider slit degrades the spectral resolution. There is a trade-off between getting more light into the spectrograph, which benefits from a wider slit, and getting the spectral resolution required to distinguish the spectral lines of interest, which benefits from a narrower slit.

Once the light is inside the spectrograph, its intensity is further diminished: the typical reflectance of each aluminized surface is only ~85% in visible light, each air–glass interface typically transmits only 96% of the light, and the diffraction of light into different spectral orders by a grating often means that only 20% to 60% of the light ends up in the spectral order you are trying to observe. (These percentages can be improved by applying special coatings to the surfaces, which may improve reflectances to 95% and transmissions to 99%.)

■ If each air–glass interface transmits only 96% of the incident intensity, what fraction of the light incident on a lens will be transmitted?

❏ A lens has two air–glass interfaces, one where the light enters and one where it exits. If the intensity of the incident light is I_0, then the fraction of light transmitted by the first surface is $0.96I_0$. At the second surface, the incident intensity is therefore $0.96I_0$, of which 96% will be transmitted, so the total transmission is $0.96 \times 0.96I_0 = 0.92I_0$. That is, such a lens transmits ~92% of the incident light.

The guidelines for a particular telescope and spectrograph should indicate how long an exposure is needed to record a certain number of photons from a star of a given brightness, according to certain assumptions about the slit width, the seeing conditions, and the zenith distance of the object under investigation. The photon count rate will also depend on the dispersion of the spectrograph set-up, i.e. how much it is spread out. Clearly there are numerous factors that affect whether the recorded spectrum will contain the number of photons needed in order to be able to distinguish the spectral lines you want to see from the noise inherent in the observations.

3.4 Other spectrographs

Although the simple, single-slit spectrograph described above is the type you are most likely to find on a small telescope, there are other more complex designs available. Each of these has its own role to play in astronomical observations.

An **echelle spectrograph** has a second dispersing element, either a second grating or a prism, which disperses the light at right angles to the direction of dispersion produced by the main grating. Without going into details, the effect is to produce a spectral image that consists of a stacked series of spectra (see Figure 3.9). Each of the stacked spectra represents a part of the spectrum of the object, spanning only a very narrow range of wavelengths. You can imagine joining these individual spectra end to end in order to assemble the complete spectrum of the object.

Figure 3.9 The spectral image produced by an echelle spectrograph. Each band comprises a small part of the spectrum covering only a very narrow range of wavelength. (The vertical streak is a fault on the detector.)

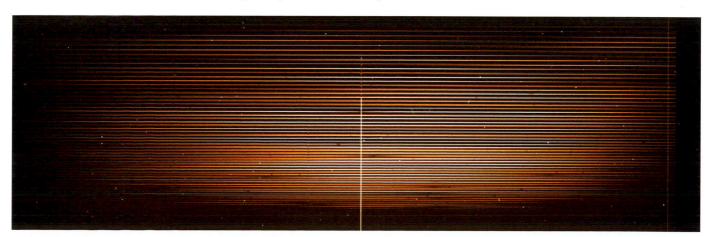

■ What do you suppose are the advantages of an echelle spectrograph? What are its disadvantages?

❏ An echelle spectrograph enables us to cover a *large* wavelength range at a *high* spectral resolution. However, since the light is dispersed over a large part of the image plane, the intensity at any point in the spectrum is very low, so these instruments can only be used successfully on very large telescopes or with bright stars.

Integral field unit spectrographs and multi-object **fibre-fed spectrographs** use optical fibres to feed light from various parts of the focal plane of the telescope through gratings to produce many individual spectra on the same detector. In an integral field unit, the fibres are closely packed together so that a spectrum from every point on a two-dimensional image of an extended object may be produced. Such instruments are useful for mapping the velocity field across a spiral galaxy, for instance. In multi-object fibre-fed spectrographs, the individual fibres may be automatically positioned at any location in the field-of-view, to feed the light of many hundreds of individual objects onto the spectrograph (see Figure 3.10). These instruments are useful for obtaining the redshifts of hundreds of galaxies within a single image, for instance. In both types of instrument, the resulting image consists of a series of individual spectra stacked one above another, essentially covering the whole of the detector.

An alternative to using a fibre-fed spectrograph is to use a **multi-slit spectrograph**. This technique is identical to single-slit spectroscopy as described earlier, except that, as implied by its name, there are multiple slits in the field-of-view. Each of these slits allows light from a different object to pass into the spectrograph and form a spectrum on the detector. In order to align the slits in their correct positions, a separate mask is usually prepared for each field to be observed with the slits simply drilled in the appropriate positions. Multi-slit spectroscopy has none of the throughput losses that are associated with passing light through optical fibres. However, a disadvantage of the method is that the field-of-view is usually smaller

(a) (b) (c)

Figure 3.10   (a) The 400 optical fibres on the 2dF (two degree field) instrument at the Anglo–Australian Telescope.
(b) A close-up of part of the field plate showing some of the fibres positioned in the field-of-view.
(*Both pictures © Anglo–Australian Observatory.*)
(c) A schematic diagram showing the head of each optical fibre, clamped accurately in position on the field plate using a strong magnet. Light from the telescope enters the microprism and then passes down the optical fibre to be dispersed by a grating and the spectrum fed onto a detector.

than for fibre spectroscopy. A modern multi-slit spectrograph may have a field-of-view that is only around 10′ in diameter. This may be compared with the 2° diameter field of the 2dF shown in Figure 3.10. Also, in order to prevent spectra from overlapping on the frame, the number of spectra which can be recorded simultaneously is usually less than 50 rather than the 400 that are possible with a device like the 2dF.

3.5 Summary of Chapter 3 and Questions

- In an astronomical spectrograph light may be dispersed using either a prism, a reflective diffraction grating or a volume phase holographic diffraction grating.

- The *grating equation* quantifies the amount by which light of different wavelengths is dispersed by a grating having a particular groove spacing.

- In a spectrograph, light is first collimated before passing through the dispersive element, and then focused by a second lens or mirror before arriving at the image plane. A slit in the telescope focal plane allows closely spaced neighbouring wavelengths in the image plane to be distinguished.

- The throughput of a spectrograph is diminished by reflection from, and absorption by, each optical element within the instrument.

- *Echelle spectrographs* use a second disperser at right angles to the first, to obtain a high spatial resolution over a large range of wavelength. *Integral field unit spectrographs* and multi-object *fibre-fed spectrographs* use optical fibres to collect light from many different parts of a telescope's field-of-view and enable spectra of many parts of a single extended object, or many individual objects, to be obtained simultaneously. *Multi-slit spectrographs* also allow the spectra of many objects to be obtained simultaneously.

QUESTION 3.1

The light from a star is incident normally on a reflective diffraction grating with 120 grooves per mm. In which orders of the spectrum does the red end of the spectrum in one order (at 700 nm) first overlap with the blue end of the spectrum in the next order (at 400 nm)?

QUESTION 3.2

(a) A spectrograph contains one lens, a mirror and a reflection grating. In the visible waveband (from about 400 nm to about 700 nm), each air–glass interface transmits 85% of the light incident on it, and 4% of the incident light is also absorbed by each reflection. What is the fraction of incident visible light from a star that emerges from the spectrograph?

(b) For the spectrograph in (a), 40% of the emergent light occurs in the first-order spectrum, spanning the range 400 nm to 700 nm. What is the intensity of light per nm of wavelength range as a fraction of the incident light from the star in this wavelength range?

4   ASTRONOMICAL DETECTORS

Human eyes respond only to the *rate* at which light from the source is reaching the retina. Once this rate falls to the threshold of sensitivity, the visibility of such a weak source cannot be improved by gazing at it for a long time. In fact, because of stress and tiredness, the very opposite tends to happen. Hence, although a telescope with a great light-gathering power undoubtedly helps to discover new weak sources, the sensitivity threshold of the eye is still a limiting factor.

4.1 Integrating detectors

If we replace the eye by an **integrating detector**, then by using long time exposures it is possible to detect sources that are several orders of magnitude weaker than can be detected with the eye. An integrating detector is so called because it can integrate (i.e. add up) the light it receives over a long period of time. The only limitation here is the background brightness of the sky itself. If the exposure is long, the detector will eventually record the intensity of the scattered light in the atmosphere, and the faintest astronomical sources will remain lost in this background. Ultimately, it is a combination of expert judgement, trial and error, and often a measure of luck, which leads to the most perfect images of the night sky.

In order to take images with long exposures, modern telescopes are equipped with sophisticated automatic guiding devices. As noted in Chapter 2, they make it possible to fix the field-of-view of the telescope on one particular object (or on a particular section of the sky) and to keep this field-of-view constant with such smoothness and precision that the images exhibit no loss of resolution, even though the Earth has been rotating around its axis and moving along its orbit during the exposure.

The importance of taking images through a telescope lies not only in the fact that images can record weaker sources than the eye can see; equally important is the fact that such images provide a permanent and accurate record of the observation.

An integrating detector is also capable of recording finer details in the structure of extended celestial objects, or of separating more closely spaced point-like objects, than can be immediately seen by the eye through the eyepiece. One reason for this is that the individual recording elements of a detector can be packed more closely than the receptors on the retina of the eye. Another reason is that the detector is often positioned in the focal plane of the primary mirror (or lens) and is therefore not affected by the residual aberrations of the eyepiece. The detector can therefore, in principle at least, make full use of the angular resolution of the telescope. Although, in practice, such limits of resolution are impossible to achieve when observing stars from the Earth's surface because of atmospheric turbulence, there is still a gap between the best angular resolution that can be achieved by the telescope and the acuity of the eye. An imaging detector can record details on scales of 1″ or better that can subsequently be enlarged to make them visible by eye.

Black-and-white photographic emulsion was the first, and still is the simplest, type of integrating detector used in connection with astronomical telescopes. However, for several reasons, photographic emulsion is *not* an ideal detector of the relative brightness of celestial bodies. Firstly, its response is non-linear. This means that the intensity recorded on the developed photograph is not directly proportional to the brightness of the light falling on it. (Although this is an inconvenience, the effect can often be allowed for and calibrated accordingly.) Secondly, and more importantly, photographic emulsions have a relatively low sensitivity to light when compared to electronic detectors. For more accurate photometric (i.e. brightness comparison) measurements of individual sources, it is preferable therefore to use some form of photoelectric detector.

4.2 CCDs

A photoelectric detector is essentially a device that responds to incoming photons of light by producing an electrical signal. This electrical signal is then detected, amplified and measured, and the resulting image is built up and processed using a computer. Several forms of photoelectric detector have been used over the years, such as *photomultipliers* and *photodiodes*, but nowadays, the commonest type of detector used in astronomy is known as a **CCD**, which stands for *charge-coupled device*.

A CCD is a two-dimensional, highly sensitive solid-state detector which can be used to generate, extremely rapidly, a pictorial representation of an area of the sky or a spectrum. Similar detectors are now routinely used in digital cameras. Figure 4.1 shows an example of an astronomical CCD. As you can see from the figure, physically CCDs are very small, typically only a couple of centimetres across. They are usually made from a silicon based semiconductor, arranged as a two-dimensional array of light-sensitive elements. The pictures generated from such detectors therefore consist of an array of picture elements, known as **pixels** for short, with one pixel in the image corresponding to each light-sensitive element in the CCD. Conventionally therefore, the light-sensitive elements of the CCD itself are also referred to as pixels.

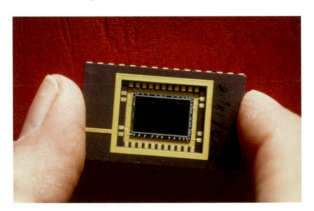

Figure 4.1 An example of a CCD used in astronomical imaging. (Courtesy of *John Walsh/ Science Photo Library.*)

Note that $4096 = 2^{12} = 4 \times 2^{10}$.

The individual pixels on the CCD can each be considered as tiny detectors in their own right. A modern CCD may contain up to 4096×4096 (referred to as $4k \times 4k$) pixels in an array, with each pixel typically of order 10 to 20 μm across. When light falls on a pixel, each photon generates one electron–hole pair in the semiconductor; the electron is called a **photoelectron**, since it is produced by a photon. Hence the number of pairs depends on the intensity of the radiation. Once an exposure is completed, the accumulated charges are transferred out of the array in a controlled manner, one row at a time. This is converted into a digital signal which can be displayed on a monitor screen and stored on computer for later processing and analysis.

CCDs have an extremely high efficiency at visible wavelengths, recording typically 70% of the photons that fall on them. This may be compared with the efficiency of photographic emulsion which is typically only a few percent.

■ A CCD consisting of 1024×1024 light-sensitive elements arranged in a square array is used to obtain an image of a star cluster under seeing conditions of 1″. In order to obtain an image that is 'well matched' to the seeing, it is reasonable to have the image of a point object stretching across about four pixels on the CCD. Hence, each pixel on the CCD must correspond to an angular size of 0.25″, to avoid an image falling on the dead area between pixels.

(a) If the physical size of the detector is 20 mm × 20 mm, what is the scale of the image formed on the CCD in arcseconds per mm?

(b) What is the field-of-view of the CCD in this case?

(c) What focal length telescope is required to match this performance?

❑ (a) 1024 pixels occupy 20 mm on the CCD, hence each is only (20 mm/1024) = 19.5 μm across. Each pixel corresponds to 0.25 arcseconds on the sky image, hence the image scale is (0.25 arcseconds/19.5 μm) = 1.28×10^4 arcseconds m^{-1} = 12.8 arcseconds mm^{-1}.

(b) The whole CCD is 20 mm across, so the field-of-view is (12.8 arcseconds mm^{-1} × 20 mm) = 256 arcseconds along each side, or 4.3 arcminutes × 4.3 arcminutes.

(c) A field-of-view spanning 256 arcseconds corresponds to (256/3600) × (π/180) radians = 1.24×10^{-3} radians. This will extend over 20 mm when the focal length of the telescope is f_o = 20 mm/1.24×10^{-3} = 16 100 mm or about 16.1 m.

CCDs are now used for virtually all astronomical imaging and the images so obtained can then be used for astrometry (measuring the positions of objects), astronomical photometry (measuring the brightness of objects) and spectroscopy (measuring the spectra of objects).

4.3 Summary of Chapter 4 and Question

● *CCDs* have many advantages over the eye as a detector for use with an astronomical telescope. They are *integrating detectors* and so can detect *fainter* objects than the eye; they also enable a *permanent* record to be kept of the observation, and they allow *finer detail* to be investigated than possible with the eye alone.

● Unlike photographic images, those produced by CCDs have high efficiency, photometric linearity and the ability for images to be processed and analysed by computer techniques.

Figure 4.2 shows a (schematic) CCD spectral image, obtained using a single-slit grating spectrograph, containing the spectra of two stars. The CCD comprises 800 × 800 pixels arranged in a square array and has a linear size of 10.0 mm along each side. The telescope used to obtain the image has an effective focal length of 4.00 m.

(a) What is the image scale in the plane of the telescope, in arcseconds per mm?

(b) What is the field-of-view at the detector in arcminutes?

(c) What is the angular scale of the image in arcsec per pixel?

(d) What is the angular separation of the two stars?

(e) If the spectral scale of the image is 0.4 nm per pixel, what is the difference in wavelength of the two emission lines (labelled H_α and H_β) in each spectrum?

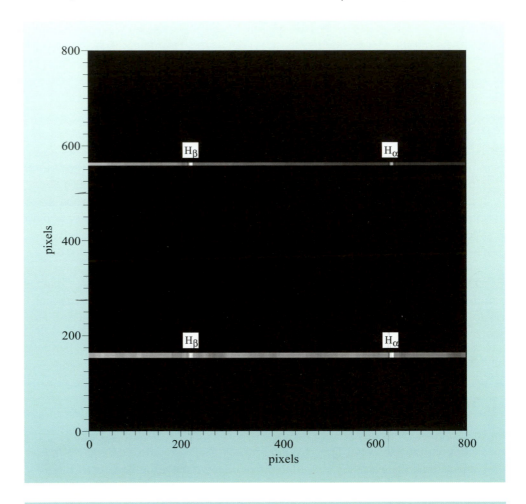

Figure 4.2 The CCD spectral image referred to in Question 4.1.

5 REDUCING CCD DATA

A CCD is just a semiconductor chip. To make it into a useful astronomical device, it must be connected up to electronics that power it, control it, and allow its data to be read out. The charge that has accumulated in each detector pixel is initially read out as a tiny electric current. This current is amplified and converted into a number expressed in so-called analogue data units (ADU). The ADU value is therefore a measure of the charge that was read out from the detector pixel in question, but is on an arbitrary scale. In order to quantify the number of photoelectrons that the pixel held, and therefore the number of photons that were incident on the pixel during the exposure, an analogue-to-digital conversion (ADC) factor is applied to the number in ADUs. The ADC factor is essentially the number of photoelectrons per ADU. The result of this process is a digital image where the value in each image pixel is the number of photons incident on the detector pixel during the exposure (subject to some calibration which we discuss below). Bear in mind however, that the CCD itself does *not* count photons – a distinction which is crucial in assessing uncertainties.

The numbers in a digital image are stored to a certain numerical precision. For instance, each pixel value may be stored as a 16-bit binary number. In this case, the maximum value that may be stored in any pixel of the image is $2^{16} - 1 = 65\,535$, i.e. there are 65 536 unique values possible, ranging from 0 to 65 535. If the light intensity falling on any single detector pixel is such that it would generate a value larger than this limit, the analogue-to-digital conversion will **saturate**. In such a case, the value stored in an image pixel would no longer be representative of the number of photons falling on the detector pixel at that location. Exposure times need to be carefully selected so that the objects of interest do not saturate the CCD. If necessary, a number of short exposures may be combined to overcome this.

■ What is the maximum value that can be stored in an image pixel produced by a CCD whose analogue-to-digital converter operates with 15-bit precision?

❑ The maximum value is $2^{15} - 1 = 32\,767$.

A raw CCD image is shown in Figure 5.1. Reducing, or processing, CCD data consists of taking the array of values stored on the pixels of an image, and manipulating it appropriately to produce an image in which the numerical value in each pixel is directly proportional to the number of photons falling on the detector at that location. In order to do this, a series of corrections need to be applied to each raw image, and we consider these below.

5.1 Bad pixels and cosmic rays

In any CCD, some of the detector pixels will be faulty and will return values that are misrepresentative of the light falling on them. Such pixels are referred to as *hot* or *cold* or *bad*. Sometimes an entire column or row of the CCD may contain **bad pixels**. Software to process CCD images will generally have the facility to either ignore bad pixels, or to replace them with an interpolated value based on the values in adjacent non-bad pixels.

Figure 5.1 A raw CCD image showing bad pixels, bad columns and varying intensity from one region to another.

When ionizing radiation, either from local, naturally occurring sources of radioactivity or from cosmic rays, hits the CCD it releases charge in a pixel that is similar to that caused by light falling on the chip, though often many times greater. These spurious signals are usually confined to a single detector pixel or a few adjacent pixels, and any individual image may have several dozen or several hundred **cosmic ray events**, with the number increasing with exposure time. However, as cosmic ray events have abnormally high values in single pixels they are usually easy to distinguish from genuine 'point sources' such as stars whose light will be spread over a few pixels with a characteristic distribution, namely the point spread function of the telescope as discussed in Chapter 2 Section 4. Automated software to remove cosmic ray events from individual images operates in the same way as for bad pixels by interpolating using adjacent pixel values. Note, however, that this is only a partial correction, since the interpolated value is only an estimate of the real one.

Alternatively, since the locations of cosmic ray events are entirely random, they may be removed by combining together several individual images of the same field, obtained with the same exposure time. By taking the **median** value in each pixel from such a stack of images, the anomalously high values in pixels affected by cosmic rays may be rejected.

■ A stack of nine CCD images contains the following values in a particular single image pixel: 4356, 4421, 4324, 4309, 4401, 4967, 4397, 4391, 4364. Which image contains the cosmic ray? What is the median value in the pixel?

❏ The sixth image (with a pixel value of 4967) contains the cosmic ray. The median value is the fifth value when the pixel values are arranged in order. The pixel values in order are 4309, 4324, 4356, 4364, 4391, 4397, 4401, 4421, 4967. So the median value is 4391.

Notice that the median value is *not* equal to the *mean* value: the mean value is much higher due to the influence of the anomalous value in the image affected by the cosmic ray.

5.2 Bias and dark-current subtraction

When the signals from a CCD are digitized in the analogue-to-digital conversion process, an offset or **bias** signal is intentionally introduced into the digital value to prevent the signal from going negative at any point (which it may otherwise do, due to random fluctuations or noise). The value of the bias may vary depending on the position on the CCD and can vary with time over the seconds or minutes it takes to read out the array. There are two techniques for correcting for this bias signal.

Overscan strips are narrow regions of the CCD image, usually running down either side of the image, a few tens of pixels wide. They are virtual pixels created by continuing to read out the CCD even after the last real pixel has been read out. That is, they do not correspond to a piece of the detector, and hence contain no photoelectrons of their own. They therefore indicate how the CCD electronics, and the analogue-to-digital converter in particular, responds to a genuine zero signal, and how that response varies with time. For each row in the CCD, the values of the signal in the overscan strip corresponding to that row may be averaged and subtracted from the value in each other pixel in that row. After this stage of processing, the overscan strips may be discarded, thus reducing the sizes of images.

(Sometimes overscan strips are called bias strips, but this can lead to confusion with the bias frames described below.)

Bias frames are entire images created by reading out the CCD following a zero second exposure, which means there are no photoelectrons stored in its pixels. (A CCD 'image' is often called a **frame** and the two terms should be seen as interchangeable.) The bias frame enables the average noise across the chip to be measured and accounted for. Bias frames are usually obtained by taking zero-length exposures with the shutter closed at the beginning and end of each night's observing. A master bias, see Figure 5.2, is generated by creating the median of a stack of many such frames, and this can then be subtracted from every other image obtained during the night.

Figure 5.2　A master bias frame showing small-scale structure in the noise across the CCD. There is an overscan strip down the extreme right-hand side.

Another effect is that some signal may be generated in CCD pixels even when no light is present. This is referred to as **dark current** and is due to the motion of electrons that arises from the thermal energy of the CCD and defects. Like the bias signal, it varies from pixel to pixel and also changes with time. Dark current can be minimized by cooling the CCD to liquid-nitrogen temperatures. However, if this is not possible, or not sufficiently effective, then dark current may be accounted for by taking long exposures with the shutter closed, removing the bias, and then dividing by the exposure time to obtain the dark current per second in each pixel. This may then be scaled by the exposure time of every other image, and subtracted off. Dark current is often insignificant for many visible-light CCDs, but is more important when working in the infrared.

5.3 Flat-fielding

The sensitivity to light of the many pixels in any CCD will vary slightly with position, by a few percent. This is due to irregularities introduced by the manufacturing process. In order to calibrate for this relative variation in pixel-to-pixel sensitivity, a CCD is exposed to a uniformly illuminated light source, such as the twilight sky, or the inside of an illuminated observatory dome. The images obtained by such a process are known as **flat fields**. Target images may then be corrected, using flat fields, to the values they would have had if all the detector pixels had the *same* sensitivity to light. This process is known as flat-fielding. It is important to note that the pixel-to-pixel variation will also be a function of wavelength, so when observing through filters, flat fields must be obtained through the same filters as the target observations.

Flat-fielding also corrects for effects such as dust particles on the CCD itself; dust on the filters which cast ring-shaped (out of focus) shadows; and the dimming of objects observed towards the edge of the telescope field-of-view (this is known as *vignetting*) caused by obstructions in the light path or just the change of angle. Two types of flat fields commonly used are as follows.

Dome flats are images of the inside of the observatory dome, illuminated by a continuum spectral source such as a tungsten-filament light bulb. The dome will necessarily be out of focus, and the images will be featureless. Dome flats may be taken during the day, before or after an observing session. They have two disadvantages though. First, light reflected from the dome enters the telescope at a different angle from that at which light from the sky enters. This can affect the response to vignetting and dust on the filters or CCD. Second, the spectrum of a tungsten-filament light bulb is not the same as that of the night sky and can make it more difficult (or even impossible) to correct for an effect known as **fringing**. Fringing is caused by interference between rays from multiple reflections, within the CCD or filters, of light at a single wavelength. It can give rise to wave-like patterns of intensity variation across the CCD, known as **fringes**. In order to correct for fringes, we must use a flat-field source whose spectrum closely matches that of the image in question.

Fringing can usually be accounted for by using **sky flats**. These are images of the sky taken in twilight, either before or after an observing session. The sky needs to be brighter than any stars that are in the field-of-view, but not bright enough to saturate the CCD. Disadvantages of sky flats are that it can be difficult to judge the appropriate exposure times and that sunlight reflected from the inside of the observatory dome can also reach the CCD, affecting the vignetting response as with dome flats.

The procedure to 'flat-field-correct' the images is that several flat fields are de-biased and dark current subtracted, and then combined (median stacked) to produce a single master flat (in each filter); see Figure 5.3. The value in each pixel of the master flat is then divided by the mean value of all the pixels. This has the effect of normalizing the mean value of the master flat to unity (i.e. a value of one). De-biased, dark-subtracted target images in each filter are then *divided by* the normalized master flat image in the appropriate filter to remove the pixel-to-pixel sensitivity variation, dust images, and vignetting effects (Figure 5.4).

Figure 5.3 A master flat field showing spurious images due to dust particles, vignetting effects and pixel-to-pixel sensitivity variations.

Figure 5.4 A de-biased, flat-fielded image of Figure 5.1. Comparison of Figures 5.1 and 5.4 as 'before' and 'after' illustrates why we say the data have been reduced.

5.4 Summary of Chapter 5 and Questions

The stages in reducing CCD data are as follows:

- Bad pixels and cosmic rays are removed by interpolating across the affected pixels or by taking the median of a stack of images.
- A master bias frame is created and subtracted from each image, or overscan strips are used to correct for the bias level in each image.
- The dark current is removed if necessary, scaled to the exposure time of each image.
- Dome flats or sky flats in each filter are combined and normalized to a mean of unity, then divided into each target image to correct for pixel-to-pixel sensitivity, dust particle shadows and vignetting.

QUESTION 5.1

Figure 5.5 shows cross-sections through the images of two stars on a CCD frame. One of the two stars is saturated. Which one is it and why? How could you avoid this problem?

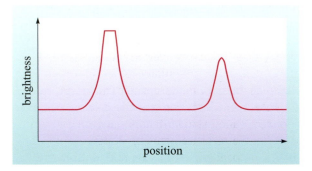

Figure 5.5   Cross-sections through the images of two stars on a CCD frame. One of the stars is saturated, the other is not. See Question 5.1.

QUESTION 5.2

If a series of exposures are made of targets through different filters, why must flat fields also be obtained through *each* filter too? Conversely why is a single set of bias frames and dark frames sufficient?

6 PHOTOMETRY

Photometry is the technique of measuring the brightness of astronomical objects. But what does it mean to talk about the 'brightness' of an astronomical object? Astronomers often use the term 'brightness' rather loosely. It can refer to the amount of light, or other radiation, *emitted* by a star or galaxy; it can refer to the amount of light *reflected* by a planet or asteroid; or it can refer to the amount of light, or other radiation, *received* from an astronomical object here on Earth. When we wish to be more specific, we shall use the term **luminosity** to refer to the *intrinsic* brightness of an astronomical object, that is the amount of energy radiated by it per second. We shall also use the term **flux** when referring to the *apparent* brightness of an astronomical object, that is the number of photons from the object received per second here on Earth.

Astronomical photometry is generally carried out using observations made through one or more filters, but may also be performed without a filter in so called 'white light'. After acquiring images at a telescope and reducing the CCD frames as described in Chapter 5, the stages involved are usually as follows. The apparent brightness of an object on a frame is determined by defining a circular zone called an **aperture** around the object of interest and measuring the amount of light that has been recorded in that aperture. The amount of light is normally expressed as an astronomical magnitude. The measurement may be recorded as the object's magnitude relative to another star in the field or its magnitude relative to an arbitrary sky value. This apparent brightness value may then be converted into a *catalogue magnitude* in one of two ways. First, if the apparent magnitude of a comparison star in the same field is known, the relative magnitude of the target star can simply be scaled relative to that. Alternatively, similar procedures can be carried out for a number of other stars of known magnitude, called standard stars, measured at various positions around the sky. Based on the measurements of standard stars, a calibration is made to take account of the atmospheric extinction and to determine the offset between the instrumental magnitudes and catalogue magnitudes in the absence of extinction (the so-called *zero point*). By applying these corrections to the measured magnitudes of the target stars, their catalogue magnitudes may be calculated. These stages are described in detail below.

6.1 Astronomical magnitudes

The flux of light received from a star is commonly expressed on a logarithmic scale, which is defined such that a difference of 5 magnitudes represents a ratio of 100 in flux or apparent brightness. So the **apparent magnitudes** m_1 and m_2 of two stars with fluxes F_1 and F_2 are related by

$$m_1 - m_2 = 2.5 \log_{10}(F_2/F_1) \tag{6.1a}$$

or $\quad m_1 - m_2 = -2.5 \log_{10}(F_1/F_2)$ (6.1b)

Clearly if $F_1 = 100F_2$, then $m_1 - m_2 = -2.5\log_{10}(100) = -2.5 \times 2 = -5$. The negative sign indicates that brighter stars have *smaller* (less positive) magnitudes. The zero chosen means that the brightest stars have an apparent visual magnitude, m_V, around -1 while the faintest stars visible to the naked eye have m_V about $+6$. Apparent magnitude for any other wavelength range can be similarly defined.

■ The bright star Rigel has an apparent visual magnitude $m_V = 0.12$, while the faint star Ross 154 has an apparent visual magnitude $m_V = 10.45$. What is the ratio of the visual flux of Rigel to that of Ross 154?

❑ Dividing each side of Equation 6.1b by -2.5 and then calculating 10 to the power of each side,

$$10^{(m_1 - m_2)/-2.5} = \frac{F_1}{F_2}$$

so, $F_1/F_2 = 10^{(0.12 - 10.45)/-2.5} = 10^{4.132} = 1.355 \times 10^4$

Therefore Rigel is visually about 13 600 times brighter than Ross 154.

The apparent magnitude of a star is *not* an intrinsic property of the star itself – it also depends on the distance to the star and the amount of intervening (interstellar) material that absorbs and scatters the light. By contrast, the **absolute magnitude** of a star (which is related to the luminosity) *is* an intrinsic property of the star itself.

The absolute magnitude (represented by M) is defined as the value of the apparent magnitude that would be obtained at the standard distance of 10 pc from a star, in the absence of any intervening matter. To see how this is related to the apparent magnitude, consider the following.

As the light from a star streams out into space, it will become spread out over the surface of an imaginary sphere of ever increasing radius, d. Since the surface area of this sphere is given by $4\pi d^2$, the flux observed from a star will be inversely proportional to the square of the distance away from the star. Hence, for a star at a distance d away, the flux is given by $F(d) \propto 1/d^2$. We may therefore write the ratio of the flux at a distance d to the flux at a distance of 10 pc as

$$\frac{F(d)}{F(10)} = \left(\frac{10}{d/\text{pc}}\right)^2$$

So, using Equation 6.1, the relationship between apparent and absolute magnitudes is therefore

$$m - M = -2.5\log_{10}\left(\frac{F(d)}{F(10)}\right) = -2.5\log_{10}\left(\frac{10}{d/\text{pc}}\right)^2$$

or $\qquad M - m = 5\log_{10}\left(\frac{10}{d/\text{pc}}\right) = 5\log_{10}(10) - 5\log_{10}(d/\text{pc})$

Consequently,

$$M = m + 5 - 5\log_{10}(d/\text{pc}) \tag{6.2a}$$

where d is the distance to the star. If intervening material is present which scatters and absorbs light from the star, this equation may be modified to take account of the reduction in flux caused by interstellar extinction as follows:

$$M = m + 5 - 5\log_{10}(d/\text{pc}) - A \tag{6.2b}$$

where A is the amount of **interstellar extinction** expressed as an equivalent number of magnitudes. (Note that interstellar extinction includes the effects of both **absorption** of light and **scattering** of light by intervening gas and dust. Both of these effects will reduce the amount of light reaching the observer.) The **distance modulus** of a star (or other astronomical object) is defined as the difference between its apparent and absolute magnitudes, hence from Equation 6.2b,

$$\text{distance modulus} \equiv m - M = 5\log_{10}(d/\text{pc}) - 5 + A \tag{6.3}$$

■ The bright star Rigel has an apparent visual magnitude $m = 0.12$ and is at a distance of $d = 280$ pc. Assuming there is negligible interstellar extinction in our line of sight to Rigel, what is its absolute visual magnitude? What is its distance modulus?

❏ Using Equation 6.2b,

$$M = 0.12 + 5 - 5 \log_{10}(280 \text{ pc/pc}) - 0 = -7.12$$

$$m - M = 0.12 - (-7.12) = 7.24$$

Absolute magnitudes are related to the luminosities (L) of stars in a similar way to that shown in Equation 6.1 for apparent magnitudes and fluxes, namely

$$M_1 - M_2 = 2.5\log_{10}(L_2/L_1) \tag{6.4a}$$

or $\quad M_1 - M_2 = -2.5\log_{10}(L_1/L_2) \tag{6.4b}$

Apparent and absolute magnitudes can be quoted in any one of several broad-band regions of the spectrum. Commonly these are expressed as U, B, V, R, and I, standing for (respectively) near ultraviolet, blue, visible (i.e. green–yellow), red, and near infrared, and are referred to as the *Johnson photometric* system. In addition, three further bands in the near infrared are referred to as J, H and K. The wavelength ranges corresponding to these regions are shown in Table 6.1. Subscripts on m or M indicate the waveband in question, alternatively, the apparent magnitudes are themselves represented by the symbols U, B, V, R, I, J, H and K.

Table 6.1 Some standard (broad-band) photometric wavebands.

Johnson system			Sloan filters		
Filter	Central wavelength	Width of band	Filter	Central wavelength	Width of band
U	365 nm	70 nm	u′	353 nm	63 nm
B	440 nm	100 nm	g′	486 nm	153 nm
V	550 nm	90 nm	r′	626 nm	134 nm
R	700 nm	220 nm	i′	767 nm	133 nm
I	900 nm	240 nm	z′	835 nm onwards	—
J	1.25 μm	0.24 μm			
H	1.65 μm	0.4 μm			
K	2.2 μm	0.6 μm			

Other filter systems are also becoming more widespread. One such system is the filter set used for the Sloan Digital Sky Survey. The filters here are named u′, g′, r′, i′ and z′ and their characteristics are also shown in Table 6.1. These filters are designed so that specific spectral line features fall within the **pass band** of certain filters. This can help to broadly characterize astronomical objects from photometric data alone without the need for spectroscopy.

Yet another filter set is the narrow-band Strömgren filter system. The filters here are labelled u (ultraviolet), v (violet), b (blue), and y (yellow) and are centred at 350 nm, 410 nm, 470 nm and 550 nm. Unlike the Johnson and Sloan filters, the Strömgren filters each have a bandwidth of only 20 to 30 nm. They too are useful for selecting particular spectral regions that correspond to specific spectral features.

6.2 Aperture photometry

Having acquired CCD images through a variety of filters, and reduced the CCD frames using the techniques described in Chapter 5, the next stage is to perform **aperture photometry** on the objects of interest. Image analysis software tools allow a circular zone called an aperture to be defined around the star or other object of interest (Figure 6.1), and another aperture is used to define the background or sky brightness. This second aperture may be on another part of the same image, or often a concentric annulus is used, as shown here. By adding up the amount of light recorded in each aperture, the software calculates the flux due to the object in question and subtracts from this the flux due to the background. Note that, if the object aperture and the background aperture have different sizes, it will be necessary to scale the background measurement to the equivalent value that would be measured in an aperture of the same area as that used for the target. However, this is usually done automatically by software. In addition to the above, it is sometimes appropriate to select a reference star on the same image (using a further pair of source and background apertures), to which the flux from the target star will be compared.

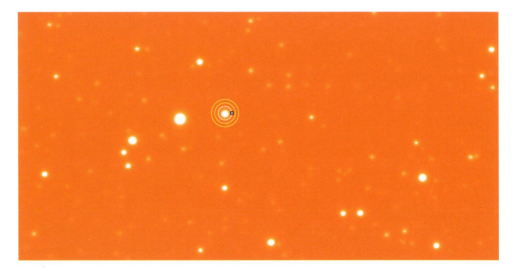

Figure 6.1   The inner circle illustrates the aperture over which the star's brightness is measured. The outer two circles denote an annulus of sky used to determine the background brightness level.

The flux from the background subtracted target is then displayed as either the magnitude relative to the reference star on the same image, or as the magnitude relative to the sky. This latter value is referred to as an **instrumental magnitude** – it is a relative value, often set such that the sky has an arbitrary magnitude of 50 (or some other suitably faint magnitude value).

You may be wondering, how does one choose the correct size of apertures for the object and the sky? This can indeed be a tricky problem. Figure 6.2 shows a cross-section through the brightness profile of the target star in Figure 6.1. Should one make the aperture radius 15 pixels or more to include *all* the light from the star? If this is done there is a danger of including excessive noise from the sky background that will dominate the outer parts of the aperture. Should one instead set the aperture radius as 5 pixels or less to include only the brightest parts of the star's image and ignore the faint extensions to its brightness profile? If this is done there is a danger of underestimating the star's apparent magnitude. Clearly there is a compromise to be achieved between including all the light from the star, and including excessive amounts of noisy sky background. Some software packages make use of **optimal photometry** routines in order to adjust the apertures to the optimum size and maximize the signal-to-noise ratio of the measurement by weighting the values of individual pixels according to their different contributions from the sky and the target. Where this is not possible, it is often appropriate to use an aperture such that all of the light from the object is included down to ~1% of the value at the peak of the profile. This will slightly underestimate the apparent magnitude of each star, but provided all stars are measured with the same sized aperture, and all stars have similar profiles irrespective of where they are in the image, the same fraction of the light will be missed within each aperture. The magnitude difference between any two stars will therefore be correct.

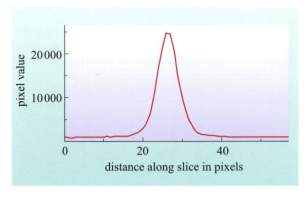

Figure 6.2 A cross-section through the brightness profile of the target star in Figure 6.1.

6.3 Magnitude calibration relative to a comparison star

The simplest type of photometry is known as **differential photometry**. When investigating the brightness variations of a variable star, for instance, you can simply monitor the *difference* in magnitude between the target of interest and another star in the same field whose magnitude is known (or assumed!) to be constant. If all you are interested in is the amplitude and timescale of variation of the target star, then it is not even necessary to know the catalogue magnitude of the comparison star. Any variations in the atmospheric extinction that affect the observed magnitude will affect both the target and comparison star in the same way. So even though the star may rise and set through the night, differential photometry relative to a star of constant magnitude in the same field will reveal the variations in the magnitude of the target star.

If catalogue magnitudes of a target object are required, this may be achieved as long as a star of known apparent magnitude is also present in the same image. **Relative photometry** is carried out by measuring the magnitude of the target star (or stars) relative to a comparison star in a similar way to the differential photometry described above. The extra step is to add on the known magnitude of the comparison star to the magnitude difference between the target(s) and comparison stars. As long as the apparent magnitude of the comparison star has been well determined previously, the magnitudes of all the stars in the same field can be determined too.

6.4 Magnitude calibration using standard stars

What do you do though if there are not any comparison stars of known magnitude in the field containing your target of interest? In this case, in order to convert the measured instrumental magnitudes into catalogue magnitudes, a magnitude calibration must be performed. The procedure is to take images of several **standard stars** through the same filters as are used for the target stars. Standard stars are objects whose apparent magnitudes have already been measured in the same wavelength range and calibrated against the (small) set of primary standard stars which *define* the magnitude system.

The Bright Star Catalogue (BSC) is a catalogue of around 9000 stars spread over the whole sky, with magnitudes brighter than $m_V \sim 6.5$. It therefore comprises all the stars visible to the naked eye. Some of the stars in this catalogue can be useful as approximately standard stars. Note though that some of the BSC stars are variable objects, and are flagged as such in the catalogue. Clearly, variable stars should *not* be used as calibrators because their apparent magnitude is not fixed at a constant value. Be careful to choose only stars of constant magnitude as standard stars!

A more reliable source of standard stars, chosen for photometric constancy, is the collection of over 500 equatorial stars in the magnitude range $11.5 < m_V < 16.0$ published by Landolt (1992). The Landolt stars have the advantage that they are grouped together into regions of the sky, such that there will be several standard stars observable in one typical CCD frame, though the stars are all much fainter than those in the BSC.

A third source of standard star magnitudes (and by far the largest) is the Tycho catalogue. This was produced by ESA's Hipparcos satellite which operated from November 1989 to March 1993. The catalogue contains two-colour (B and V) photometry of over a million stars and is a complete record of the entire sky down to $m_V \sim 10.5$. For stars brighter than $m_V \sim 9$, the magnitudes are accurate to about 0.01 and are obtained from over 100 measurements per object. This large resource is accessible on the internet, for instance at the Centre de Données astronomiques de Strasbourg *http://cdsweb.u-strasbg.fr/*

In the simplest terms, there are two factors to allow for when calibrating magnitudes. The first is the **atmospheric extinction**, due to absorption and scattering of light by the Earth's atmosphere, which will vary depending on the zenith angle at which an object is observed, on the wavelength of observation, from night to night and from location to location. Objects at the zenith will have minimum atmospheric extinction, while those close to the horizon will have maximum atmospheric extinction. The second calibration is to determine the offset between the measured instrumental magnitude and the catalogue magnitude in the case of zero atmospheric extinction. Both these corrections may be carried out simultaneously if a suitable set of standard star observations are made. The key is to observe standard stars at a *range* of zenith angles *throughout* the night, using the *same* set-up as used for your targets.

For each standard star observed through a particular filter, you should measure its instrumental magnitude and record its **airmass**. The airmass is the ratio of the thickness of the atmosphere at the observing altitude to the thickness at the zenith. It is therefore a dimensionless number. If we make the assumption that the Earth's atmosphere consists of plane-parallel layers then, as shown in Figure 6.3, the ratio of the thickness of the atmosphere at the zenith to the thickness of the atmosphere at a particular observing altitude, is simply given by cos(*zen*). Hence the airmass may be approximated as:

$$\text{airmass} = 1/\cos(zen) \tag{6.5}$$

We then assume that the following equation applies:

$$(m' - m) = \varepsilon X + \zeta \tag{6.6}$$

where m is the catalogue magnitude of the star (also known as the true apparent magnitude), m' is the instrumental magnitude of the star, ε (the Greek letter epsilon) is the **extinction coefficient** in magnitudes per airmass, X is the airmass and ζ (the Greek letter zeta) is the **zero-point offset** between instrumental and catalogue

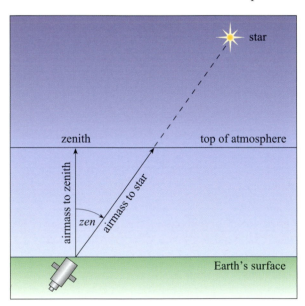

Figure 6.3 Assuming a plane-parallel atmosphere, the ratio of the thickness of the atmosphere at the zenith to the thickness of the atmosphere at a particular observing altitude, is simply given by cos(*zen*).

magnitudes. Hence if we plot a graph of $(m' - m)$ versus X, the standard star measurements through a particular filter should all lie along a straight line of gradient ε and intercept ζ. In order to determine these constants, you may choose to follow just one standard star over the night, and measure its brightness at a range of zenith angles as it rises and then sets. However, it is usually best to use a small set of standard stars so that some can be measured at large zenith angles and others at small zenith angles at any given time. Note also that the extinction coefficient will usually vary from night to night, so must be re-determined for each night's observations. If you are unlucky, it may also vary significantly throughout a single night's observing.

See Chapter 15 on graph plotting if you are unfamiliar with the equation of a straight line graph.

Note that in practice, the terms zero point and zero-point offset are interchangeable.

■ What is the airmass of a star at the zenith? What is the airmass of a star at an elevation of 30° above the horizon?

❏ A star at the zenith has a zenith angle of 0°, so its airmass is $1/\cos(0°) = 1.0$. A star at an elevation of 30° above the horizon has a zenith angle of $(90° - 30°) = 60°$, so its airmass is $1/\cos(60°) = 2.0$.

Having determined both the extinction coefficient and zero-point offset for each filter, the catalogue magnitudes of the target objects are determined by applying Equation 6.6 once again. With measured values of instrumental magnitude and airmass, and calculated values for ε and ζ, the catalogue magnitudes of the target objects may be calculated in each filter.

Note that if you are observing targets that are always close to the zenith (say, zenith angle <10°), then the extinction coefficient will be of negligible importance in your calibration. Only the zero-point offset will be relevant. In this case you can get away with only measuring standard stars close to the zenith too and then assuming that the difference between catalogue magnitude and instrumental magnitude is roughly constant for both your standard stars and your targets.

The calibration procedure described above uses two parameters, the extinction coefficient ε and the zero-point offset ζ, to convert the instrumental magnitudes to standard magnitudes. For greater precision, two more parameters may be needed: the **transformation coefficient** and the **secondary extinction coefficient**. Although it is unlikely you will need to use them to analyse observations made for training purposes, you would need them for accurate research, and for that reason we briefly introduce them here. The transformation coefficient adjusts for differences between the equipment you use and the equipment used by whoever made the standard measurements, e.g. Landolt. Differences in the way the transmissions of filters and the sensitivities of detectors vary with wavelength are adjusted for by the transformation coefficient. Secondly the extinction coefficient ε treats all of the wavelengths transmitted by a given filter as being extinguished equally, but this is an over-simplification since stars observed at high airmass not only appear fainter but also redder – as you may have noticed if you have ever watched the sunset. The secondary extinction coefficient allows for this second fact.

6.5 Summary of Chapter 6 and Questions

• The fluxes of light from stars may be characterized by their *apparent magnitude* through one or more filters. The magnitude scale is logarithmic, and the *brighter* the star, the *smaller* the numerical value of the magnitude.

- *Aperture photometry* is carried out on reduced CCD images by measuring the flux within a circular aperture around the object of interest and subtracting the sky flux. The magnitude may be determined relative to that of a comparison star in the same image, or relative to that of the sky.

- *Differential photometry* involves simply measuring the difference between the flux of the (variable) target star and a (constant) comparison star on the same image. *Relative photometry* of one or more stars in an image may be achieved by measuring their flux relative to a star of known magnitude contained in the same image.

- Measurements of *standard stars* at a range of *zenith angles* allow the *extinction coefficient* and *zero-point offset* to be determined for each filter. Instrumental magnitudes of target stars may be converted into catalogue magnitudes by applying a calibration determined using standard star observations.

QUESTION 6.1

The apparent visual magnitudes of Rigel and Ross 154 are 0.12 and 10.45 respectively, while their distances are 280 pc and 2.9 pc respectively. What is the ratio of the V-band luminosities of Rigel and Ross 154? (Assume that there is negligible interstellar extinction along the line of sight to each of these stars.)

QUESTION 6.2

An aperture of radius 12.0 pixels is placed around a star on a CCD image and encloses a total count of 2.50×10^6 photons. An annulus of inner radius 18.0 pixels and outer radius 24.0 pixels surrounds the star and includes a total background count of 7.00×10^5 photons. Assigning an arbitrary instrumental magnitude of $m_b = 50.0$ to the background sky flux in 1 pixel, what is the instrumental magnitude of the star?

[NB Remember to correct the sky flux to the same aperture area as the target star before doing the background subtraction.]

QUESTION 6.3

The star Regulus ($m_V = 1.35$, $m_B = 1.24$) is used as a standard star and observed throughout the night through a V-band filter and a B-band filter. Its instrumental magnitudes measured by aperture photometry at a range of airmasses are as given in Table 6.2. (Remember that the instrumental magnitude is on an arbitrary scale.)

(a) What are the zero point and extinction coefficient in each of the V-band and B-band?

(b) A target star was measured to have an instrumental V-band magnitude of 29.75 and an instrumental B-band magnitude of 31.25, at a zenith angle of 35°. What are the catalogue magnitudes of the target star?

Table 6.2 For Question 6.3.

zenith angle	airmass	V-band instrumental magnitude	B-band instrumental magnitude
60°	2.00	24.32	23.09
40°	1.31	24.19	22.91
20°	1.06	24.17	22.82
40°	1.31	24.23	22.85
55°	1.74	24.28	23.01
65°	2.37	24.38	23.20

7 SPECTROSCOPY

As explained in Chapter 3, the spectral images produced by a single-slit grating spectrograph generally consist of a band of varying intensity stretching across the image in the *spectral* direction, often with different spectra separated vertically on the image in the *spatial* direction.

- ■ If the spectral direction of the image produced by a single-slit spectrograph runs *horizontally* on a CCD, in which direction must the slit have been aligned?

- ❏ The slit must have been aligned perpendicular to this direction, parallel to the spatial direction of the image on the CCD, i.e. vertically.

Note that in the following we shall always assume that the spatial direction runs vertically on the CCD frame and that the spectral direction runs horizontally, with the shortest wavelength on the left. If individual frames acquired on a particular telescope do not match this arrangement, it is possible to flip them left-to-right and/or rotate them through 90° using a software image analysis package. We now consider how to extract spectra from such spectral images.

7.1 Flat-fielding CCD frames containing spectra

We described earlier how to flat-field correct CCD frames that are simple images. There is one difference when dealing with CCD frames containing spectra, and that is that one dimension of the CCD image now corresponds to a spectral direction rather than a spatial direction. A conventional flat-field correction of the type described earlier is therefore *not* appropriate for spectral images. This is because the variation in the sensitivity of the CCD along the spectral direction is wavelength dependent, and the spectrum of the light-source illuminating the flat field is *not* the same as the spectrum of the target objects. It is essential that the flat-field illumination for spectroscopy should contain no emission lines, so, for example, fluorescent lights cannot be used.

The approach in this case is therefore as follows. The values in the pixels in each *column* of the master flat field image are added together to obtain a single 'row' of data. The values in this row of pixels are divided by the number of pixels originally present in each column to give the average signal in each column. A smooth curve is fitted to these data to model the spectral response of the spectrograph to the light-source used in the flat field. The value of the master flat field is then divided at each point by the value of the smooth curve to produce a new flat field that now only contains the spatial variation in the CCD's sensitivity (i.e. in the vertical direction), provided the flat-field illumination was uniform along the slit. The spectral image values are then divided by this modified flat field as before in order to correct for pixel-to-pixel sensitivity in the vertical (spatial) direction only.

7.2 Extraction of spectra

In simplest terms, the extraction of a spectrum from a spectral image involves identifying the range of rows of the CCD that encompass the spectrum, and then summing the pixels along each short column of the CCD over these rows to produce

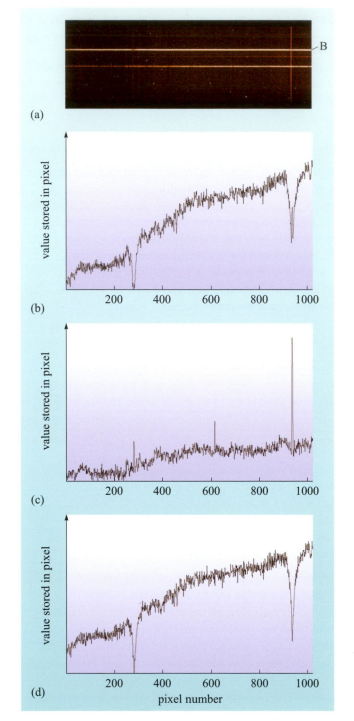

Figure 7.1 (a) A region of the spectral image is selected encompassing the spectrum of an object of interest. As noted in Figure 3.8, this spectral image contains the spectra of two bright stars and three faint ones. The spectrum of one of the bright stars (star B) is selected in this case. (b) The rows of data corresponding to the selected object are averaged to produce the spectrum shown here. (c) Rows of data corresponding to the background sky are averaged to produce the spectrum shown here. (d) The result of subtracting the sky spectrum from the target spectrum.

a single row (see Figure 7.1a and b). One or more regions of background or sky spectrum are selected from the image in a similar way and subtracted from the target spectrum (see Figure 7.1c and d), allowing for the possibility that a different number of rows may be used in the extraction of the target and background spectra.

In the example shown here, the resulting sky-subtracted spectrum of the target object shows two broad absorption lines superimposed on a smoothly varying continuum. Notice that the emission line present in the sky spectrum around pixel 940 appears in the 'raw' target spectrum superimposed on the absorption line, but has been removed by the simple sky subtraction procedure. The variation in the continuum of the resulting sky subtracted spectrum is determined by the underlying spectral distribution of the object *and* by the spectral response of the CCD. Furthermore, the spectral axis of this graph is still in terms of pixel number. The next stages in calibrating such a spectrum involve assigning wavelength values to the horizontal axis and accounting for the spectral response of the detector.

7.3 Wavelength calibration

In order to assign wavelength values to the horizontal axis of a spectrum such as that shown in Figure 7.1d, the spectrograph is usually exposed to a standard emission line source. Usually this is a hollow cathode lamp, commonly, but incorrectly, referred to by astronomers as an **arc lamp**. This is a discharge tube filled with a vapour whose spectral lines are at well determined wavelengths (see Figure 7.2a). The lamp spectra are extracted in a similar way to target spectra, except there is no need of a sky subtraction, as the exposure is all internal to the telescope and the light from the lamp fills the slit. The pattern of emission lines, such as that in Figure 7.2b is compared with a standard list of lines that are known to be produced by the vapour in question. In this way the wavelengths corresponding to a

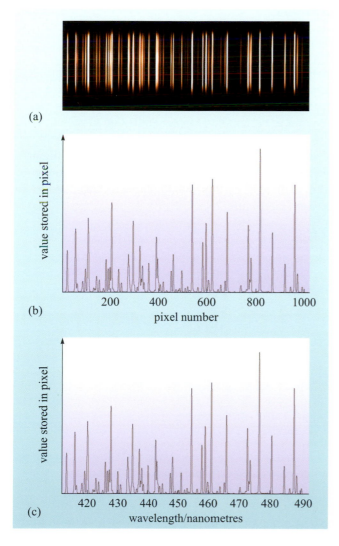

Figure 7.2 (a) A spectral image obtained by exposing a CCD to a spectrum from a copper–argon arc lamp. (b) An arc spectrum obtained from a copper–argon arc lamp with the horizontal axis in pixels. (c) The same spectrum with the axis now calibrated in nanometres.

variety of pixel locations may be established. A simple mathematical shape, such as a cubic function, is fitted to establish the wavelength corresponding to *any* pixel position, and this is applied to the lamp spectrum (Figure 7.2c). The wavelength axis is then simply copied to the target spectrum (Figure 7.3).

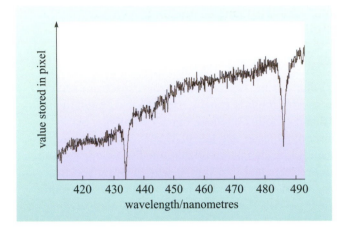

Figure 7.3 The sky subtracted spectrum with the axis now calibrated in wavelength.

7.4 Flux calibration

As mentioned earlier, a spectrum such as that represented by Figure 7.3 still contains the effect of the wavelength response of the CCD. Therefore the variation in strength of the continuum seen there does not necessarily reflect the true variation in the continuum spectrum of the source. In many cases we are only interested in determining the presence or absence of certain spectral lines, or of measuring the accurate wavelengths of certain lines to determine Doppler shifts due to radial velocity motions, for instance. In this case a full flux calibration is not important, and the simplest thing is simply to remove the wavelength response by normalizing the continuum to a mean level of unity. To do this, a smooth curve is fitted to the continuum and then the spectrum is divided by this smooth curve to create a **continuum normalized spectrum** (see Figure 7.4).

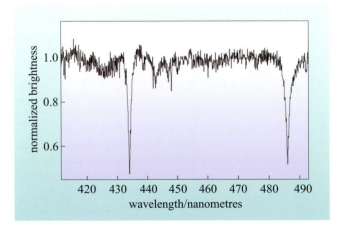

Figure 7.4 The continuum normalized spectrum created by fitting a smooth curve to the continuum trend in Figure 7.3, and then dividing the spectrum by this trend.

In cases where a full flux calibration of the spectrum is required, it is necessary to obtain spectra of spectral standard stars whose flux is known at a number of wavelengths in the range that is being observed. There exist catalogues of spectral standard stars that have been measured in just this way, such as those by

Oke & Gunn (1983) and by Filippenko & Greenstein (1984). The technique is then to compare your observed spectra of spectral standards with the published flux calibrated spectra, and determine the instrumental response as a function of wavelength (position on the CCD). The final step is then to divide target spectra by this instrumental response to end up with a flux calibrated spectrum of your target. The details of this procedure have only been outlined here, as it is unlikely that as a student you will carry out a full-flux calibration of this kind on your data.

7.5 Displaying spectra

Spectral information is usually presented in the form of a graph. The horizontal axis of the graph usually shows wavelength or frequency. (In spectral regions other than the optical, the horizontal axis may show some other equivalent quantity, such as photon energy for γ-ray and X-ray spectra, or the number of wavelengths that fits into 1 cm of length, a quantity called the **wavenumber** which is commonly used by infrared astronomers.) The vertical axis of the graph normally indicates the brightness of the spectrum at any given wavelength, and again a variety of equivalent quantities can be used. In its simplest form the vertical axis may record the number of photons that were measured from a source during a particular exposure. However, as convenient as this may seem, it tells the reader nothing directly about the intrinsic or even apparent brightness of the source. The vertical axes of some graphical spectra are labelled **spectral flux density**. At any given wavelength, λ, the spectral flux density, $F_\lambda(\lambda)$, expresses the amount of energy per second that passes through a unit area (1 m^2 in SI units) per unit interval of wavelength $\delta\lambda$, where $\delta\lambda$ is a narrow wavelength interval centred on λ. The λ in the subscript reminds you that F is defined per unit wavelength interval, and the λ in brackets reminds you that F varies with wavelength. In SI units, this might have the unit W m^{-2} nm^{-1}. A similar, but distinctly different, physical quantity that may also be plotted, and which is regrettably also called the spectral flux density, is the quantity given the symbol $F_\nu(\nu)$.

- By analogy with the definition of $F_\lambda(\lambda)$, can you guess what the definition of $F_\nu(\nu)$ might be?

- $F_\nu(\nu)$ expresses the amount of energy per second that passes through a unit area (1 m^2 in SI units) per unit interval of frequency $\delta\nu$, where $\delta\nu$ is a narrow frequency interval centred on ν. The ν in the subscript reminds you that F is defined per unit frequency interval, and the ν in brackets reminds you that F varies with frequency. In SI units, this will have the unit W m^{-2} Hz^{-1}.

Graphical spectra may also use vertical axes that show **relative spectral flux density** where the spectral flux density at any wavelength is expressed as a *fraction* of some reference value and there will be no SI units shown on the axis. A special case of value for stars is where the reference spectrum is the continuum level of the same spectrum, which is what the spectrum would look like in the absence of any discrete spectral lines or bands. As noted earlier, a spectrum which has been divided by its own continuum is said to have been continuum normalized, so much of the spectrum will have a value of 1.0 on the vertical axis, with absorption and emission lines falling below or rising above that level.

7.6 Summary of Chapter 7 and Question

- Spectral images have to be flat-field corrected in a different way from regular images because the variation in the horizontal (spectral) direction is due to the varying spectral response of the spectrograph and the spectral distribution of the lamp used to make the flat-field image.

- Spectra are extracted by summing over a number of rows in the image and then subtracting a similarly extracted sky spectrum.

- Wavelength calibration is carried out by matching the pattern of emission lines seen in the spectrum of an arc lamp to the known wavelengths expected from the vapour in question.

- Flux calibration in its simplest terms can involve normalizing the spectrum to a smooth fit to the continuum in order to remove the effect of the varying wavelength response of the spectrograph and the continuum spectral distribution of the source. Alternatively, if a genuine flux calibration is required, spectral flux standard stars must be observed.

- Spectra are conveniently displayed with a vertical axis corresponding to *spectral flux density* $F_\lambda(\lambda)$. This is the amount of energy per second that passes through a unit area per unit interval of wavelength.

QUESTION 7.1

Explain the difference between the horizontal stripes apparent in the spectral image illustrated in Figure 3.9 and the vertical stripes apparent in the spectral image illustrated in Figure 7.2a.

8 MICROSCOPES AND MICROSCOPY TECHNIQUES

Not all aspects of astronomy and planetary science are confined to observing objects at vast distances. We are fortunate in having a supply of solid samples available for study in the form of **meteorites** and **interplanetary dust particles** as well as extraterrestrial samples brought back from space missions. All of these samples are made of rocky material and therefore many of the techniques employed in the study of geology can be readily applied to planetary samples. The rarity of such samples, as well as their random delivery and collection, means that the study of their relationships on a large scale is not possible. In contrast, on Earth one might study how the rock types change across a mountain range. Instead, the laboratory study of planetary materials is largely confined to the details *within* each sample – rarely larger than a few tens of centimetres across and generally much smaller, down to fractions of a millimetre in the case of dust particles. While present day research into planetary materials is conducted with a vast array of highly sophisticated analytical tools, the initial characterization of each sample and selection of suitable portions for further study is still performed by optical examination using a **petrographic microscope** (or polarizing microscope) (Figure 8.1).

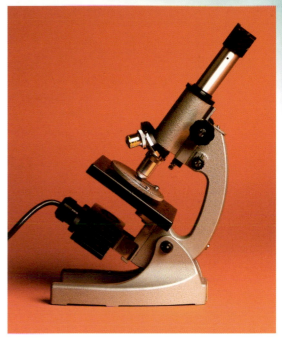

Figure 8.1   A simple petrographic microscope (or polarizing microscope).

Microscopes have been around since the seventeenth century, but the development of what we now know as the petrographic microscope was not made until the mid-nineteenth century by Henry C. Sorby in Sheffield. Much criticized at the time, Sorby countered that

> '…no one expected astronomers to confine their observations to what they could see with the naked eye, so why should geologists be so restricted'

and perhaps more pertinently noted that even when studying such features as mountains, or for that matter planetary bodies,

> '…there is no necessary connection between the size of an object observed and the value of the facts and conclusions that can be derived from it.'

The study of rocks with a microscope requires very careful preparation of the sample, typically no more than 2 or 3 cm across, into a very thin wafer, usually about 30 microns thick, mounted on a glass slide and polished to produce a high quality finish. Even with today's automated machines, this is a time consuming process and it takes several days to prepare a section.

The key element of the petrographic microscope is the use of polarized white light to view the sample. Although most minerals in rocks are largely transparent when only 30 microns thick, it is the interaction of polarized light with the minerals that offers enhanced contrast and therefore improved image quality compared to normal microscopes. The use of polarized light also permits use of a rather surprising optical property of minerals which is of great help in their identification.

8.1 The interaction of light with minerals

The interaction of light with the rock-forming minerals in thin section lies at the heart of the unique properties of the petrographic microscope. It is outside the scope of this book to explain these complex phenomena in detail but the fundamental principles are outlined here.

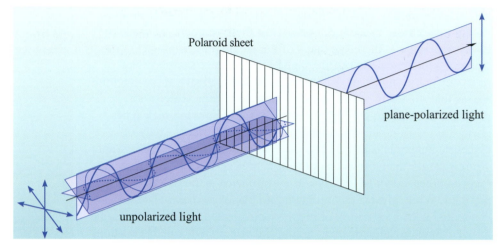

Figure 8.2 The production of plane-polarized light by passing unpolarized light through a polarizing filter, in this case a Polaroid sheet.

A single ray of light (or indeed any electromagnetic radiation) is composed of two oscillating fields at 90° to each other and to the direction of propagation: namely an electric field and a magnetic field. For simplicity we will concentrate on the electric field. The plane containing the oscillating electric field is called the **plane of polarization** of the ray. Usually, a beam of white light will contain many such rays all randomly rotated with respect to each other. However, the polarizing filter in a petrographic microscope only permits light of one plane of polarization to be transmitted, and the light so produced is called **plane-polarized light** (see Figure 8.2).

The strong internal electric field associated with the atoms which make up a crystal interacts with the oscillating electric field of light passing through the crystal. One of the major effects of this is to slow down the passage of light through a crystal. (This is exactly the same phenomenon referred to in Chapter 3 in the discussion of the role of a glass prism in a spectrograph.) As noted earlier, the phenomenon of light changing speed when travelling from one medium to another is known as *refraction*. The effect is quantified by the **refractive index** of the material, which is defined as the speed of light in a vacuum divided by the speed of light in the crystal.

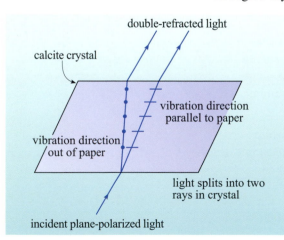

Figure 8.3 Side view through mineral showing how light is refracted into two rays as it passes through the crystal.

Crystals have very ordered arrangements of atoms. Some crystals are isotropic in that their crystallographic structure appears the same irrespective of which of any three mutually perpendicular axes it is viewed from. However, most crystals are said to be anisotropic, in that their apparent crystallographic arrangement is orientation-dependent. In such minerals, the strength of the internal electric field encountered by a ray of light is dependent on the alignment or orientation of the crystal relative to the plane of polarization of the incoming light. This results in a variation of the refractive index with the orientation of the plane of polarization. Most importantly, anisotropic minerals also have a rather remarkable property in that when plane-polarized light enters such a crystal it becomes split into two rays (for reasons beyond the scope of this book), in a process of double refraction. Both rays are plane polarized, but polarized at 90° to each other (see Figure 8.3), which means that the two rays will encounter different atomic arrangements, and so different refractive indices, and therefore travel through the crystal at different speeds. This behaviour of having two different orientation-related refractive indices is called **birefringence**.

As the two rays pass through the crystal at different speeds they will emerge from the crystal with a **phase difference** between them. This phase difference ϕ can be related to the birefringence of the crystal as:

$$\phi = \frac{d}{\lambda(e_1 - e_2)} \times 360° \tag{8.1}$$

where d is the thickness of the crystal, $(e_1 - e_2)$ is a measure of the birefringence of the crystal, i.e. the difference in the two refractive indices, and λ is the wavelength of the light.

To obtain a measure of this phase difference it is necessary to combine the two emergent rays. This is achieved by passing the two rays through a second polarizer (the analyser) which is orientated at 90° to the first polarizer. However, in almost all cases neither of the two emerging rays is parallel to the analyser vibration direction and therefore only a component of each ray can be transmitted. The components of each ray emerging from the analyser now lie in the same plane of polarization. They can be combined by a lens. If the phase difference between the two components is 180° they exactly cancel each other out and no light is transmitted.

- For a fixed sample thickness and given birefringence value, what property of the light will affect the phase difference between the two emergent rays?

- The wavelength of the light (See Equation 8.1).

Therefore, when using a white-light source, different parts of the spectrum are cancelled out for any given birefringence value and mineral thickness. The combined rays therefore generate a specific colour, which depends on the mineral under study (see Figure 8.4). With increasing birefringence the interference colour changes from yellow through to pink and the pattern is repeated as the resultant phase difference continues to increase past 360°. Therefore, by simply viewing the interference colour of a mineral in thin section and comparing the image to a calibrated standard colour chart of interference colours, it is possible to accurately and quickly determine the birefringence of the mineral, which is of great benefit in the identification process. Note, though that, this does require some skill in selecting suitably orientated minerals and determining to which cycle a particular interference colour belongs.

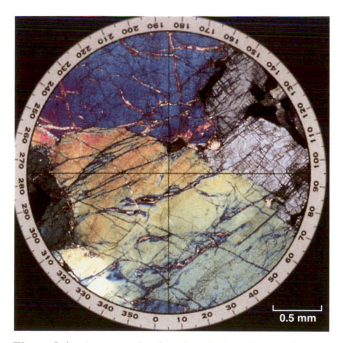

Figure 8.4　An example of a mineral viewed through polarizing filters in a petrographic microscope.

8.2 Components of a petrographic microscope

In its simplest form, a microscope consists of only two optical components, an objective and an eyepiece, separated by the body tube (Figure 8.5). The objective projects a magnified image of an illuminated object into the body tube and the eyepiece further magnifies the image projected by the objective. Examples of such microscopes have been around since the early 1600s and while the overall principle of such a two-component, or compound microscope, is at the core of almost all modern microscopes, each component is now vastly more complex and additional features are introduced into the light path.

The main components in such a microscope are:

- **Light source** – typically this is a high-energy tungsten–halogen bulb (50 to 100 watts). The DC power supply for the bulb is normally controlled by a potentiometer built into the microscope in order to control the level of illumination of the sample.

- **Polarizing filter** – a sheet of polarizing film, often mounted in a rotating, but lockable mounting, situated between the light source and the sample, to provide light having an electric field oscillation that is confined to a single plane, i.e. plane-polarized light. Conventionally this is orientated across the microscope.

- **Substage light condenser** – this usually consists of an adjustable aperture diaphragm and a series of lenses that gathers light from the microscope light source and concentrates it into a cone of light that illuminates the viewable area of the specimen with uniform intensity.

- **Sample stage** – this is a 360° rotatable stage with a mechanism to hold the sample securely in place. Focusing of the image of the sample is usually achieved by moving the sample stage towards or away from the objective lens using two geared handles to achieve coarse and fine control.

- **Objective lens** – this is perhaps the most important, and critically designed component of a modern microscope. The simple convex lens employed in the early microscopes has now been replaced with a compound lens containing up to six or more components in order to project highly magnified images that are free from chromatic and spherical aberrations. It is critically important that objective lenses are stress-free and strain-free as stresses and strains can generate the same sort of optical effects as those that are to be studied in the samples. This involves selecting special strain-free glass or minerals from which to fabricate the components, as well as ensuring that, when multiple lens elements are cemented together and mounted in tight fitting frames, no strain is imparted to the optical components. Such strain can also occur as the result of damage through dropping or even rough handling of the objective. The optical properties of any objective are fixed and therefore a number of objectives are usually mounted on a rotating turret offering a range in magnifications – typically from ×4 to ×40. The objective is focused close to the specimen in order to project a magnified real image up into the body tube of the microscope.

- **Analyser** – this is a second polarizing filter and can be inserted in the optical path at almost any point above the sample, although it is most commonly found immediately above the objective lens. This filter is fitted on a rotating mounting but should always be positioned such that its plane of polarization is at right angles to that of the lower polarizer and in such a way that it can be readily inserted and removed from the optical path. When inserted in the optical path, the sample is said to be viewed in **cross-polars**. If no sample is present in the

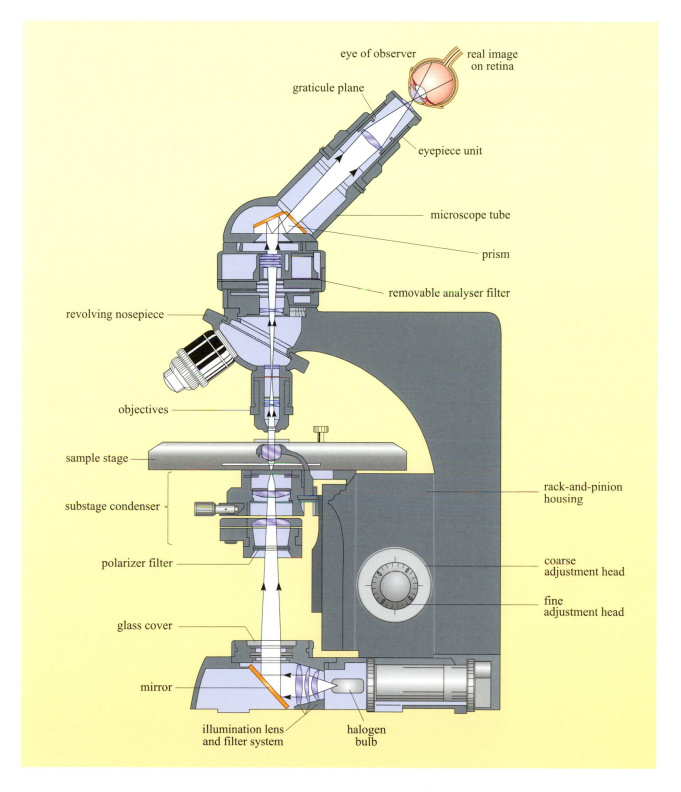

eye of observer

real image on retina

graticule plane

eyepiece unit

microscope tube

prism

removable analyser filter

revolving nosepiece

objectives

sample stage

substage condenser

polarizer filter

glass cover

mirror

rack-and-pinion housing

coarse adjustment head

fine adjustment head

illumination lens and filter system

halogen bulb

optical path then, as the two polarizing filters are at 90° to each other, none of the plane-polarized light transmitted by the polarizer will be transmitted by the analyser and the image will appear entirely dark. In the absence of a sample, anything other than a dark image indicates that the polarizer is not perpendicular to the analyser.

Figure 8.5 Schematic of a petrographic microscope.

- **Eyepiece lens** – this is the final optical component and it further magnifies the real image projected by the objective. For visual observation the eyepiece produces a virtual image which appears as if it were near the base of the microscope. As with the objective lenses, the eyepieces are also composed of a series of lenses and are often designed to work in conjunction with specific objectives to compensate for any colour-dependent differences in the magnification of the objective. Typical magnification factors for eyepieces are ×10 to ×15. Modern research microscopes are generally fitted with binocular heads which use a series of prisms to deliver the image to both eyepieces simultaneously. Individual eyepieces can also be fitted with measuring scales and crosshairs (usually referred to as **graticules**) for positioning and measurement of the specimen.

8.3 Magnification and resolution

Normally the viewing medium is air where $n = 1$, although viewing media with higher refractive indices (such as oils) can also be used with special objectives.

The magnification of a microscope is simply the product of the magnification of the objective and the eyepiece, e.g. a ×40 objective with a ×10 eyepiece gives a total magnification of ×400. While care must be taken to select eyepieces that are compatible with the objectives in terms of appropriate corrections, there is also a limit to which the eyepieces can enhance the magnification of the system. This is determined by the performance of the objective – primarily a parameter called the **numerical aperture** (NA) – which is a measure of a microscope objective's ability to gather light and resolve fine detail in the specimen. It is defined as:

$$NA = n \sin \alpha \tag{8.2}$$

where n is the refractive index of the viewing medium and α is one-half the angle subtended by the objective aperture from the position of the sample. In practical terms of manufacture, α is restricted to less than ~72° and therefore when viewing in air the maximum value of NA is ~0.95. The useful magnification that can be achieved with any lens is approximately $1000NA$, so with a ×40 objective there is no point in using eyepieces with a magnification of more than ×25, as to do so would lead to no increase in detail.

■ If a sample is mounted in oil with a refractive index of 1.1, what is the maximum useful magnification to use in an objective when coupled with an eyepiece that provides a magnification of ×15?

❏ Since the maximum value of α is ~72°, the maximum value of the numerical aperture in this case is $NA \sim 1.1 \times \sin 72° \sim 1.05$. The maximum useful magnification is therefore ~ $1000 \times 1.05 \sim 1050$. So with a ×15 eyepiece, the maximum useful magnification of an objective would be $1050/15 \sim \times70$.

Perhaps more important than magnification is the spatial **resolution** of an objective, which is defined as the minimum distance between two points on a sample which can still be resolved as distinct entities (although this is somewhat subjective especially at high magnification where an image can appear not sharp but with features resolved). The spatial resolution (R) of an objective in a microscope is related to the numerical aperture by:

$$R = \frac{0.61\lambda}{NA} \tag{8.3}$$

where λ is the wavelength of the light. (This is related to Equation 2.5 and the limit of angular resolution of an astronomical telescope.) As white light is normally used to view samples it is usual to work with a wavelength of around 550 nm (green) as this is the light to which our eyes are most sensitive. Therefore, the maximum resolution of any objective viewing the sample in air (i.e. $NA < 0.95$) is about 350 nm or 0.35 microns. Needless to say that to acquire such resolution requires the microscope to be set up perfectly with all other components suitably matched to the objective and therefore in reality the working resolution is always significantly poorer than this.

8.4 Reflected-light microscopy

Not all minerals are transparent in thin section. Some, particularly sulphides and non-silicate oxides, are essentially opaque. Therefore, for some studies it is also necessary to include the capability to view the sample with a reflected, or incident light source. This is particularly true when working with most meteorites as they tend to include significant quantities of iron–nickel metal and sulphides, in proportions from a few percent to 100%. For reflected light viewing, a second light source is required to illuminate the specimen directly along the viewing axis. After passing through a polarizing filter the light from this second source is introduced into the viewing path usually by means of a coated-glass plate reflector mounted in the viewing path at 45° and located below the analyser. Only some part of the illuminator light is actually reflected onto the specimen by the reflector, the remainder is lost from the system. Equally, only some of the light reflected off the specimen actually passes through the glass reflector plate up towards the eyepiece. The efficiency of the glass plate is such that, at most, only 25% of illuminator light reaching the reflector can actually be transmitted to the eyepiece.

The quality of the surface of the sample is critical when viewing in reflected light. The sample must be well polished with very fine abrasive paste (usually less than 1 micron grit size) and free of tarnish and other stains. To this end most planetary materials are prepared as polished thin sections which permit viewing in both transmitted and reflected light. However, this is not always necessary, for instance when a sample is largely composed of non-transmitting minerals such as an iron meteorite containing mostly iron-nickel metal. In this case polished blocks may be used, but then great care must be taken to ensure that the sample surface is perpendicular to the viewing axis. This is achieved by either carefully machining the samples so that the top and bottom faces are flat and parallel or by using simple mechanical levelling devices that press the sample block into a small lump of modelling clay on a larger glass slide.

8.5 Photomicrography

The textures of rocks as viewed in thin section through a microscope can be very complex and therefore it is useful to record these images. When learning about the mineralogy of rocks and microscopic techniques it is generally very informative for students to make detailed hand-drawn sketches of representative areas of a rock as seen through the microscope in order to ensure that they become familiar with the important features. However, once those skills are established a more rapid and more accurate method is to record the views with a camera. Previously this has been done using film cameras, however, most of this work is now done with CCD cameras where it is possible for the operator to ascertain immediately the quality of the captured image (e.g. sharpness of focus, exposure, area of interest).

For simplicity of operation it is important that the image seen by the camera should be approximately the same size as that seen through the eyepieces – an additional graticule can be added to the eyepiece to show the field-of-view in the camera. However, the virtual image viewed through the eyepiece cannot be captured by film or CCD and therefore it is necessary to redirect the optical path by means of a retractable prism or mirror to a third 'eyepiece' which projects a real image onto the CCD or film. Ideally the number of pixels of the CCD has to be adequate to ensure that no optical information is lost compared to the viewed image through the eyepiece.

8.6 Summary of Chapter 8 and Questions

- *Petrographic microscopes* used to study rock samples in thin section use *plane-polarized light* produced using a *polarizing filter*. This plane-polarized illumination can be delivered from below the samples in *transmitted light* mode or from above in *reflected light* mode.

- The magnification achievable by a petrographic microscope depends on the magnification of the *objective lens* and the *eyepiece lens* and typically ranges up to ×400 with ×10 eyepieces, or to ×600 with ×15 eyepieces. The maximum resolution of the microscope is governed by the *numerical aperture* of the objective and if perfectly set up it should be able to resolve features approximately 0.3 microns apart.

- Anisotropic minerals cause an incident *plane-polarized light* beam to be split into two beams with planes of polarization at 90° to each other (*birefringence*). The two beams travel along paths with different *refractive indices*, resulting in two light beams emerging from the mineral that are *out of phase* with each other. When the two beams are recombined using a second polarizing filter known as an *analyser* they generate characteristic *interference colours* which may be used to identify the mineral under study.

QUESTION 8.1

The human eye is sensitive to a range of wavelengths from approximately 350 nm to 700 nm. How might the resolution of a microscope be improved if this were the most important viewing consideration?

QUESTION 8.2

In no more than 200 of your own words, summarize how cross-polars are used in a petrographic microscope in transmitted light mode when examining minerals that exhibit birefringence.

9 INTERPRETING IMAGES OF PLANETARY SURFACES

So far in this book we have considered images of astronomical objects that are obtained from a huge distance away using ground-based telescopes, and also images of extraterrestrial samples obtained using petrographic microscopes. In this section we turn to images of planetary surfaces obtained by spacecraft in orbit around, or passing relatively close by, planets in the Solar System.

9.1 Clues from illumination

Many people have trouble visualizing the topography in pictures of planetary surfaces. Sometimes, what is really a hill appears to be a hole (and vice versa) and what is really a ridge appears to be a valley (and vice versa). The following advice is designed to help you overcome such optical illusions.

When you look at a picture of a planetary surface, shadows give the best clues as to the topography. If the caption tells you from which way the illumination is coming you should be able to distinguish between hills and holes by working out where the shadow should be in each case. Figure 9.2 (overleaf) is an enlargement of part of Figure 9.1, with labels to indicate some of the topographic clues. For example, the labelled peaks near the centre of the large crater cast shadows on their lower left sides. You are told that sunlight is coming from the upper right, so these have to be hills rather than holes.

Figure 9.1 Impact craters on the Moon photographed obliquely from the Apollo 11 spacecraft in lunar orbit. Sunlight is coming from the upper right. The largest crater is 80 km across. You should be able to satisfy yourself that the floor of the crater is lower than the terrain beyond the rim. The smallest craters visible in the foreground are about 100 m across. (© *NASA*)

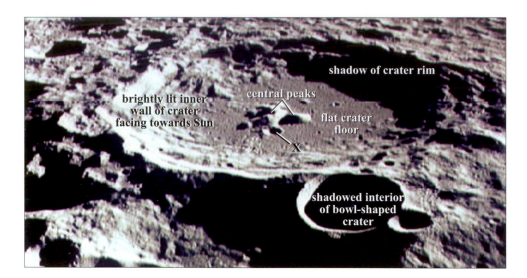

The labels on the figure read:
- shadow of crater rim
- central peaks
- brightly lit inner wall of crater facing towards Sun
- flat crater floor
- X
- shadowed interior of bowl-shaped crater

Figure 9.2 An enlargement of part of Figure 9.1. (© *NASA*)

The craters in Figure 9.1 and Figure 9.2 appear elliptical merely because the view is oblique. Vertical views show the craters to be circular.

■ Look carefully at the feature near the central peaks labelled X in Figure 9.2. Part of this is bright (sunlit) and the other part of it is dark (in shadow). Bearing in mind which way the sunlight is coming from, try to deduce whether this is a hill or a hole.

❑ This feature is brightly lit on its left side and shadowed on its right side. As the sunlight is coming from the right and yet it is the right-hand side that is shadowed, this feature cannot be a hill like the nearby central peaks. On the contrary, it must be a hole. In fact it is a small crater, which must have been formed by a smaller impact at an unknown time after the large crater was formed.

If, when you look at this picture, you perceive the topography the right way round, then that's fine. However, if you see it 'inverted' (i.e. with hills as holes) try turning the page round (90° at a time) until the topography looks correct. The human visual system expects shadows to be below rather than above an object, and in the case of Figure 9.1, an anticlockwise rotation of 90° might make things fall into perspective for you.

Just like the images discussed previously in this book, images of planetary surfaces obtained using spacecraft are invariably digital images, i.e. they are recorded electronically onto a CCD rather than on photographic film. In the case of an image of a planetary surface, the distance across the area of ground covered by a pixel is often spoken of as representing the resolution of the image. A high resolution image shows fine detail by virtue of its small pixels. Each pixel in a low resolution image covers a larger area and so the image does not show so much detail.

Normally, when you look at a digital image you do not notice the individual pixels. However, if a portion of an image has been greatly magnified, as in the example on the right of Figure 9.3, then the pixels become apparent. If you look at this from across the room, it may seem unremarkable, but as you bring it closer to your eyes its appearance will break up into a mass of little grey squares.

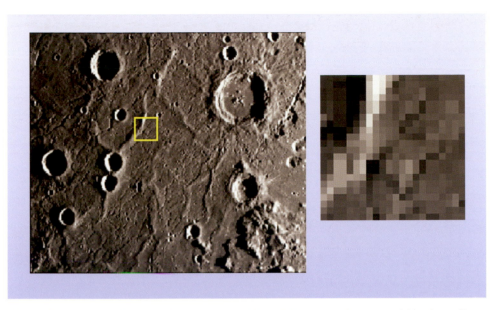

Figure 9.3 A digital image of part of the surface of Mercury. The area within the yellow box is shown enlarged on the right. This magnification reveals the individual pixels of which the image is composed. (© *NASA*)

9.2 Summary of Chapter 9 and Question

- Shadows give the best clues to the topography in planetary images. Shadows on the same side of a feature as the direction of illumination can indicate craters or valleys, while shadows on the opposite side to the direction of illumination can indicate mountains or ridges.

There are countless images of planetary surfaces obtained by spacecraft that are available on the World Wide Web. If you would like to practice your skills at interpreting such images, take a look at those available on the following sites.

General archives of NASA mission images:

> *http://photojournal.jpl.nasa.gov*
>
> *http://wwwflag.wr.usgs.gov/USGSFlag/Space/wall/wall.html*
>
> *http://pds.jpl.nasa.gov/planets/*

The Nine Planets (multimedia tour of the Solar System):

> *http://seds.lpl.arizona.edu/billa/tnp/*

QUESTION 9.1

(a) Which way is the sunlight coming from in Figure 9.3? (b) Is the long feature that cuts through the top left part of the yellow box in Figure 9.3 a ridge or a valley?

PART II: SKILLS
10   TEAMWORK

Professional astronomers and planetary scientists rarely work on their own, but more often operate as part of a team. These teams can vary considerably in size, depending on the type of work they are doing. For example, a collaboration building a facility for very high energy γ-ray astronomy in Namibia in southern Africa consists of approximately 100 astronomers, physicists and engineers. On the other hand, a team of astronomers specializing in making observations and interpreting data from a particular star or galaxy may be much smaller, and only comprise two or three people. At the same time this small team will have to work with many other people who operate the facilities that they use (such as a large optical telescope at an overseas site or an X-ray astronomy satellite). As a student, you will generally be working in small teams to make your astronomical and planetary science observations. Like professional astronomers, your team will have to work together, under a shortage of time, to successfully plan your observations, to acquire, reduce, analyse and interpret the data, and to write up or present your results.

Being able to work efficiently in a team is a vital professional skill for an observational astronomer or planetary scientist as it is in many other walks of life. The observational projects you carry out will usually be designed to be done by a team of people; there will generally not be enough observation time or facilities for you each to make a complete set of individual observations, so efficient teamwork will be essential to achieve successful project outcomes.

Planning is a vital part of observational astronomy and planetary science, and this is even more vital when working in a team, since all the members need to be clear about what tasks need doing and who will be doing what. A significant part of your project will thus involve meeting to plan the next set of observations, and, equally important, to review how the previous set of observations went.

Your group will need people to perform a number of roles for it to work efficiently. One efficient way of working is as follows. For meetings you will need a *chairperson* who should set the agenda for the meeting, steer the group through the agenda with an awareness of timekeeping, and encourage contributions from all members of the team. At the meeting, you will also need a *secretary*, who will take accurate notes (or minutes) of the meeting and so clarify what actions are to be taken. You will also need a *project manager*, whose role is to monitor the progress of the whole project against the agreed project plan and so identify any potential problem areas as early as possible and then initiate remedial action.

Normally the planning process will involve breaking the project up into a number of tasks which are then allocated to group members. In a research situation, these tasks will probably be divided up to reflect each member's individual expertise. However, in a teaching situation it is normally best if the tasks are rotated so that all students have the opportunity to develop and practise each of the roles. In allocating tasks you should consider the following:

- workloads should be fair and approximately equal,
- the work should be divided up in the most efficient way,

- each team member's personal interests should be allowed for,
- each team member's expertise should be taken account of.

Sometimes it may not be possible to initially agree such a division. In such a case it is probably better to redefine the tasks and renegotiate rather than randomly assigning tasks, which may lead to friction and lack of commitment.

It is quite likely that circumstances will change during the course of a night's observing. For instance, equipment may fail; the weather may change and so the type of observation that can be made will also change. You may need to convene a short meeting during the course of an observation to reassign tasks and priorities.

Review of progress is as important as planning. After each observing session you should meet as a group and review progress. If you are making night-time observations, then it may be best to leave this until the next day when everyone is fresh. Such a review will obviously include how well your data acquisition is going (for instance, is it of adequate quality and quantity? are you on schedule?) but should also include a review of the group effectiveness (for instance, is all the work that has been agreed actually getting done?).

One commonly-used analysis of group dynamics identifies four stages in group formation:

Forming – where the members may experience considerable anxiety as the group explores the situation and how they will interact with each other.

Storming – a period of conflict, possibly between sub-groups or individuals, in which opinions may polarize.

Norming – a group identity begins as the group coalesces; conflicts are resolved and mutual support develops.

Performing – the group adapts itself to efficiently perform the common task. Members fill functional and flexible roles.

It may be useful to be aware of these recognized stages when reviewing how your group is functioning; but remember, like most generalizations, they will not occur in all situations.

Making astronomical observations at night can create its own particular problems for group working. You are unlikely to have enough time for your body clock to totally adjust to working astronomer's hours and consequently you will not be your best at 3 a.m. At times like this, mistakes can be made, tempers get frayed and group coherence can suffer. It is very important that you understand the difficulties in working under these conditions and make allowances for them.

11 PREPARING FOR PRACTICAL WORK IN ASTRONOMY AND PLANETARY SCIENCE

An essential part of the preparatory stage of astronomical observations is to make sure that you are clear about the goals of the project. Are you aiming to measure the value of something, confirm or deduce a relationship between different variables, or simply observe a particular phenomenon? It will be helpful to read any background information that is provided about the theory that underpins the phenomenon or about the observational techniques and measuring instruments that will be used.

You may also need to do a bit of research before you start making observations. For example, you may need to prepare finding charts of the objects you are to study; and you may need to locate previously published information about your targets. We consider each of these aspects below.

11.1 Planning

Clearly you need to consider what measurements or observations you will make, and under what conditions. As an example, suppose that you were asked to investigate how the astronomical colours of the stars in a particular cluster varied with their astronomical magnitudes. You would have to decide:

- Do you need to cool down the detector before you start, do you need to set up the pointing and the focus of the telescope, and how much time should be allocated for each of these tasks?

- What calibration data such as flat fields, dark frames and bias frames will you need, and at what intervals through the night should they be obtained?

- Do you need observations of standard stars and how should these observations be distributed throughout the night, across the sky and across stars of different magnitude or spectral type?

- Which star cluster will you observe? Which ones are visible from your location at a particular time of night at a particular time of year?

- How many stars in the cluster will you take measurements of and over what range of magnitudes?

- How many different filters are needed, if any, and which particular ones should be chosen?

- What exposure times will you need in each filter, and how many observations in each filter should you take?

The answers to these questions will be governed largely by the answer to a much broader, but even more important question: what is the scientific question motivating your observations? For the example given above you should ask 'What am I hoping to learn by investigating the colour-magnitude diagram of the cluster?'

Another issue to think about is the *precision* required in the experimental measurements: for example, do you need to measure the brightness of a star to the nearest magnitude, to the nearest one-tenth of a magnitude or to the nearest one-hundredth of a magnitude? This may determine your choice of instrument or measuring technique, and it will affect the number of measurements that you make. It is also important to think about how the uncertainties in the individual measurements contribute to the overall uncertainty in the final result. This will allow you to concentrate your efforts on improving those measurements that make the largest contribution to the overall uncertainty.

You should also consider how you will analyse and display the data that you collect. What tables do you need to draw up for recording your data? Do you need to leave space for extra columns in data tables to compute other quantities that are derived from the data? What graphs might you need to plot as you are collecting data?

Before starting any practical work in astronomy and planetary science, you should consider any potential safety hazards and the precautions that can be taken to reduce risks. Health and safety legislation places responsibilities on *everyone* who works at an observatory or laboratory. The people supervising work in these facilities are required to carry out risk assessments for using the equipment, and they have a responsibility to inform students about the hazards and about the precautions that should be taken. Students then have a responsibility for taking the necessary precautions.

A final reason for spending some time on the planning stage is that you will have a better idea of what is involved in the observations, and this should allow you to divide your time appropriately between the different stages.

11.2 Astronomical finding charts and previously published data

Most modern telescopes are equipped with sophisticated pointing software, so that the observer simply has to input the name or coordinates of the object to be observed, and the telescope then automatically points to that position on the sky. However, sometimes the telescope pointing is not accurate enough to go directly to the object you want, and even if it does, you may not immediately recognize the object you want to study within the field-of-view. For this reason it is often useful to prepare **finding charts** of the objects of interest before beginning your observations. A finding chart is an image of a small patch of sky (typically several arcminutes across) centred on the object of interest. You may also use the finding chart to identify other nearby objects, their magnitudes, and positions. It is usually best to make finding charts that cover a somewhat larger area of sky than the field-of-view of the instrument you are using. In this way, with a finding chart in front of you, it is usually possible to quickly work out where your telescope is pointing, and by how much you need to slew it in a certain direction to centre on your target of interest.

One way of making finding charts is to use an on-line version of the **digitized sky survey**. The whole sky has been surveyed photographically by the Oschin Schmidt Telescope at Palomar Observatory (Northern Hemisphere) and the UK Schmidt Telescope at Siding Spring Observatory (Southern Hemisphere). The photographic plates obtained by these telescopes have been scanned to produce a digitized map of the whole sky. The digitized sky survey is available on the World Wide Web at the following addresses (amongst others):

http://ledas-www.star.le.ac.uk/DSSimage/   (at Leicester University, UK)

http://stdatu.stsci.edu/dss/   (at Space Telescope Science Institute, USA)

At these websites you can simply type in the name of the object you are studying, or its coordinates, and display the appropriate patch of the digitized sky survey. The resulting image can be printed or saved electronically for further manipulation.

The above is all very well for stars and galaxies, as they essentially remain at fixed coordinates on the sky. However, objects in the Solar System (planets, asteroids, comets, etc.) can change position significantly from night to night or in some cases over the course of a few hours. For such objects it is necessary to know precisely *where* they will be at the time of your observations. A useful website in this regard is one run by NASA's Jet Propulsion Laboratory. It is known as an *ephemeris generator* for Solar System objects, and it allows you to select your object of interest and output a list of its positions as a function of time. The website may be found at:

http://ssd.jpl.nasa.gov/cgi-bin/eph   (at Jet Propulsion Laboratory, USA)

It can also be useful to investigate previously published information on the objects of interest. A particularly good website in this regard is that of the Centre de Données astronomiques de Strasbourg (CDS). The three principal components of this site are:

Simbad – an astronomical database providing basic data, cross-identifications and bibliographic information for over three million astronomical objects.

VizieR – an interface to around 4000 astronomical catalogues and the data they contain.

Aladin – an interactive software sky atlas allowing the user to visualize digitized images of any part of the sky, and to superimpose entries from astronomical catalogues.

The three parts are all integrated. By typing in the name or position of an object of interest, you can retrieve a huge amount of previously published data. The site also allows finding charts to be prepared and overlaid with positional information from a variety of catalogues. The CDS website may be found at the following address:

http://cdsweb.u-strasbg.fr/   (at University of Strasbourg, France)

Although the CDS includes all types of astronomical object, it is most complete for stars. For research involving galaxies, the website of the NASA/IPAC Extragalactic Database (NED) is likely to be more useful. Like the CDS, this includes cross-references and bibliographic information, as well as catalogue data. It currently includes almost 11 million names for over 7 million objects, plus over 2 million references to over 50 thousand papers, and contains over 20 million photometric measurements. The NED website may be found at the following address:

http://nedwww.ipac.caltech.edu/   (at California Institute of Technology, USA)

Finally, there is a huge bibliographic database of astronomical journal articles called the NASA Astrophysics Data System (ADS) maintained at Harvard University. This contains more than 3.5 million bibliographic records, which are searchable through a query form. Using this you can retrieve abstracts and full-text scans of most of the astronomical literature published over the last century. The ADS website may be found at the following addresses:

http://adswww.harvard.edu/   (at Harvard University, USA)

http://ukads.nottingham.ac.uk/   (at Nottingham University, UK – mirror)

11.3 Summary of Chapter 11

- The key stages in preparing for practical work in astronomy and planetary science are: identify the goal of the project; study the appropriate background information; specify the target, equipment needed and details of the data that need to be obtained; consider the relevant measurement uncertainties; think about how the data will be analysed and presented; and check the relevant health and safety protocols.

- There is a wealth of astronomical data available on the World Wide Web. This includes digitized sky surveys, catalogues of positions, brightnesses and other measurements for astronomical objects, and archives of published journal papers.

12 KEEPING RECORDS

Astronomers may spend months, years or even decades on research projects. During that time, they will plan observations, make measurements and observations, and analyse and interpret data. There may be many separate parts to a project, and a complex project may involve a large team, so that information needs to be shared. There will be times when weather conditions prevent observations being made, or when equipment breaks down, when a variety of different approaches to analysing the data are tried, or when parts of the investigation are unsuccessful so that different approaches have to be devised. Imagine trying to do all of this without keeping good records of what has been done!

In addition, astronomers have to write reports, publish papers and give presentations at conferences, and this may be done months or even years after observational work was carried out. It is therefore essential to have an accessible, reliable and complete record of the work on which these communications can be based.

Learning to keep an observatory notebook is therefore an essential skill for astronomers. However, the skills that you develop by maintaining such a notebook are transferable to many other contexts. Software developers need to document the code that they develop, otherwise they will spend hours trying to remember what each part of a program does when they have to modify or update it. Journalists need to keep records of interviews and sources of information so that they can revisit them if necessary at some future time.

Using an observatory notebook will have benefits for you in the shorter term too. As a student, you may well be asked to produce a report of a project that you have done. This will be much easier to do if you have a written record of the important information in an easily located place. You may get different results from other students for a project, and want to compare the records in your notebooks to see how these differences have arisen. Or you may do a project that involves similar techniques or similar data analysis to one that you have done previously; if you can refer back to records in your notebook you may avoid having to start again from scratch.

The notes provided for student projects usually give instructions – sometimes in a fair amount of detail – on how to set up the apparatus, what data must be collected, which graphs should be plotted, and so on. This rather prescriptive approach is usually dictated by various constraints on time, place and available equipment. It means that you are very often freed from having to make detailed notes of procedure, since there is clearly little point in writing down in your notebook information that can easily be referred to elsewhere. However, you should keep a careful record of *your* observations and data, and make notes of specific details that are not described in any written instructions you have been given.

There are few rigid rules about how to keep a record of your observatory work, but we can provide the following guidelines and suggestions.

(a) Use a bound notebook *or* a ring-binder. Any graphs or images that you produce on computer or draw by hand, and any computer printouts of data, should be stuck or stapled into a notebook, or filed into a ring-binder, as soon as they are done. Always record your comments and data directly in your notebook, or on pages in a ring-binder. Resist the temptation to jot down bits of information on odd sheets of loose paper with the intention of 'neatly' copying it out later. This is an incredible waste of time, and it brings the risk of errors in copying. Also, don't be tempted to record your data in pencil with the intention of going over the data in ink later 'when you are sure that it's correct'. There will always be the temptation to erase data, and this is very bad practice.

(b) Make your record clear and concise. Your observatory notebook should constitute a complete and clear record of your work. It should contain the information needed to produce a report (or scientific paper), which you might not begin to write up until months after carrying out the observations. It should provide sufficient detail for you, or for someone else, to repeat the measurements or the data analysis. This means that it must be clearly laid out, and self-explanatory. Cryptic comments, or data without headings or units, may seem adequate at the time you record them, but will very soon become meaningless. However, you should make your notes reasonably concise; there is no need for complete sentences where a short phrase can convey the necessary information.

(c) Record all relevant information. A bold title and date at the start of the record will make it easy to locate. If you then make a brief note of the main aims of the observations, you will have an accessible reminder of what the work was about when you refer back to it.

In Chapter 11 you were advised to spend some time at the start planning how to make measurements, how many measurements to make, and so on. Your plans, and comments explaining the rationale behind them, should be recorded in your notebook. Don't worry that your plans may change – you can explain why they change in your notebook too. If you do some preliminary observations to help this planning process, note down the results and the lessons you draw from them. In particular, make a note of any changes to the way you plan to do the main observations as a consequence of the preliminary measurements, or note that the preliminary observations confirmed that your initial plans seemed appropriate.

There is no need to repeat information that is provided in any written instructions you have been given, but you should note the conditions prevailing during the measurements, and any modifications you make to the suggested procedure. Note, also, identifying details of the equipment that you use, since if your results turn out to be anomalous you might need to check for faults. You should record any special precautions that you take and any checks on procedures and measurements. Odd, unexpected, or plain inexplicable observations may turn out to be of crucial significance, so be sure to make a note of any such events or results. Write down all information, *as you go along*. Most of it will be irretrievable once you have left the observatory.

(d) Record details of observations. When taking a sequence of telescope observations, for example, it is good practice to keep a detailed log of all the exposures that you make. Often it is appropriate to fill these in on pre-printed log sheets such as that in Figure 12.1, which can be clipped into a ring-binder. Alternatively you may wish to simply tabulate the information for yourself in your notebook. The column headings shown here are the typical parameters that you might record for a given set of observations, but may vary according to the specific observations you are carrying out. If you are unsure about how to record table headings, read Box 12.1.

Date..28th September 2003.... Site.......majorca........ Photometry/Spectroscopy....Photometry...

Telescope...meade 12" LX200.. CCD camera......SBIG ST-8.. Spectrograph..............n/a..........

Notes.......Observers: A. Smith & B. Jones; weather: no cloud, temperature ~ 6°C, humidity ~ 25%........

File No.	Time/UT	Target	RA/h m s	Dec/° ′ ″	Filter or Slit width	Exposure/s	Airmass	Comments
128	21.34	bias	–	–	–	0	–	last bias frame
129	21.36	m39	21 32	+48 26	B	120	1.04	open cluster
130	21.39	m39	21 32	+48 26	V	90	1.03	open cluster
131	21.43	Lan 112	20 42	+00 19	B	60	1.23	Standards, Landolt field 112
132	21.45	Lan 112	20 42	+00 19	V	45	1.24	Standards, Landolt field 112

(e) **Space out your entries**. Space out the entries in your notebook so that you can come back and insert additional comments or additional information in the most logical and convenient place. It is generally a good idea to start the record of each new set of observations at the top of a new page.

(f) **Correcting errors in your notebook**. If you discover an error in your notebook, simply cross out the original record and insert the new one beside it, with a note (or footnote) explaining the reason for the correction. Having a record of places where you have made mistakes can be excellent for learning. Also, it is possible that you will realize subsequently that a result that you thought was erroneous is actually evidence for an important effect that you had been unaware of. There have been several occasions in the history of science when data were ignored or discarded because they were thought to be erroneous or anomalous, only to be rediscovered later by another scientist and shown to be evidence for a new effect.

(g) **Data analysis**. Data handling is covered in later sections of this book. However, bear in mind that all of the steps of your data analysis and calculation should be recorded in your notebook. Your calculations will be easier to follow when you return to them later if you lay them out neatly, with plenty of space between steps, and if you insert comments explaining the steps.

(h) **Conclusions and critical reflections**. Your final conclusions and any comments about the interpretation of your results should be noted. If appropriate, you should comment on how well your result agrees with published data or theoretical results, and suggest explanations for any differences. Also, record any thoughts that you have about how successful the project was, what its limitations were, how it might be improved and what you might do differently if you repeated the observations. Reflecting critically on your work in this way will enable you to improve your skills as an astronomer.

Figure 12.1 A typical observing log from part of a night's observing session.

BOX 12.1   UNITS IN TABLES, GRAPHS AND EQUATIONS

There is a good deal of confusion about how the unit of a physical quantity should be represented in table headings, on graph axes, or when the value of that physical quantity is substituted into an equation. Yet the international convention used in this book is both straightforward and, once the basic principles have been understood, quite logical. The most important rule to remember, and from which all the rest follow, is that whenever a symbol is used to represent a physical variable – say t to represent time, or d to represent distance – then that symbol is deemed to incorporate a numerical value *and the physical unit attached to that value*. By making this assumption, it then becomes possible to write identities such as

   $t = 5$ seconds or $d = 10$ metres

since the units balance on both sides of the equation. (If t only represented the *number*, we would have to write t seconds = 5 seconds which is something we never do!)

It now follows that, if we divide both sides of the first equation by seconds, and both sides of the second equation by metres, we have

$$\frac{t}{\text{seconds}} = 5 \quad \text{and} \quad \frac{d}{\text{metres}} = 10$$

where the quantities on both sides of these equations now have no units. And this is really how you should interpret table headings and graph axes like those shown in Figure 12.2a. By dividing the variable by the unit, the entries in the table, or the quantities on the graph are pure numbers. So point A on the graph in Figure 12.2b is at t/seconds = 5, i.e. at $t = 5$ seconds. By all means read this 'slash' notation as meaning 't in the unit of seconds' if you find this easier, but always keep in mind that the 'slash' really does represent a true division.

Occasionally, tables are drawn up as in Figure 12.3. This format is still logical – the table reads directly: distance = 10 m, or time = 15 s, etc., but notice that the unit must now be included with *every* entry in the table. However, the format in Figure 12.2a is generally neater, simpler and minimizes the amount that you have to write.

Substituting numerical values into equations follows exactly the same logic as above. Suppose you have the equation

   $d = vt,$

where d represents distance, v represents speed, and t represents time. The first point to note is that the units on either side of this equation are balanced, because the symbols are deemed to incorporate the units. But now suppose you are asked to calculate the distance travelled when the speed is 2 metres per second, and the time is 10 seconds. All you need to remember is that when you substitute the values, *you must substitute the units as well*; only in this way will you keep the units on either side of the equation balanced. So, substituting for the speed, you can write

   $d = (2 \text{ m s}^{-1})t,$

and then substituting for the time, you can write

   $d = 2 \text{ m s}^{-1} \times 10 \text{ s}$

or $d = 20 \text{ m}$

where the s has cancelled the s^{-1}.

You will see that we always include units in our numerical working in this book, particularly in the Answers to the Questions, and you should be careful to do this in your own numerical calculations.

distance	time
metres	seconds
0	0
10	5
20	10
30	15

(a)   (b)

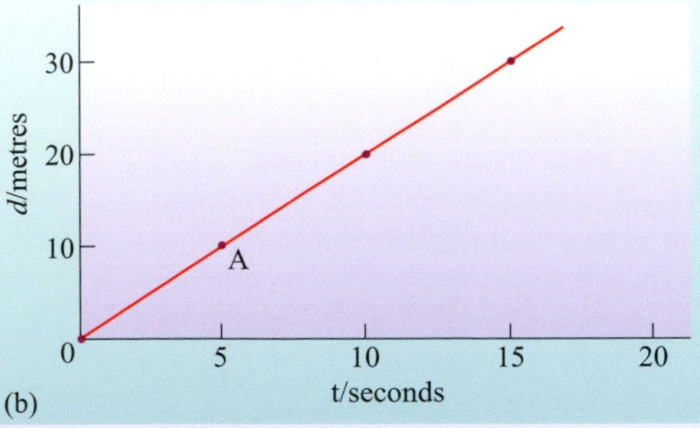

Figure 12.2   (a) Units in table headings; (b) units on graph axes.

distance	time
0 m	0 s
10 m	5 s
20 m	10 s
30 m	15 s

Figure 12.3   A less common, though still logical, way of completing a table.

13 EXPERIMENTAL UNCERTAINTIES

Measured values of physical quantities are never exact. There are always **uncertainties** associated with measurements, and it is important to assess the size of the uncertainties and to quote them alongside the measured values. So if astronomers carried out some observations to determine the apparent magnitude of a particular star, then the form in which they would quote their result would be $m_V = 12.3 \pm 0.2$. This means that their best estimate of the value is a V-band magnitude of 12.3, and their confidence in this value is quantified by the uncertainty ± 0.2, that is, the true value is probably between 12.5 and 12.1.

The value of the uncertainty conveys important information about a result, as you can see by considering the following questions.

- ■ Two astronomers make measurements of the magnitude of a particular star at the same time but using different telescopes and detectors. One quotes the result as $m_V = 12.3 \pm 0.1$ and the other as $m_V = 12.6 \pm 0.3$. In which result would you have more confidence?

- ❏ The quoted uncertainty in the result of the first observation is one-third of that obtained in the second observation. This indicates that the first observation was carried out more carefully, or used better equipment, or used a better technique, so it would be reasonable to have more confidence in the first result. (This assumes, of course, that the quoted uncertainties are realistic!)

- ■ The value of the magnitude of a particular star is measured on two different occasions by the same astronomer, using the same technique. She finds that the first time the value is $m_V = 12.1 \pm 0.2$ and that the second time the value is $m_V = 12.3 \pm 0.2$. Do these results indicate that the apparent magnitude of the star is different at the two times?

- ❏ The answer is: not necessarily. Her results are consistent, for example, with the magnitude of the star being $m_V = 12.2$ on both occasions, since this value falls within the uncertainty ranges for both observations. (The difference between the two values might motivate the astronomer to devise a more precise measuring technique. Had the uncertainties been ± 0.01, then the results would provide very strong evidence that the star's magnitude was different on the two occasions, since the uncertainty ranges would have been clearly separated.)

- ■ An astrophysicist uses a mathematical model of the star to *calculate* a value for its magnitude. His result is $m_V = 12.2 \pm 0.1$, whereas the observationally *measured* value of the star's magnitude is $m_V = 12.15 \pm 0.02$. Is the result for the star's magnitude predicted by the model consistent with the observational measurement?

- ❏ Yes, the prediction and the measurement are consistent, since the uncertainty ranges overlap.

These three examples illustrate the importance of attaching an uncertainty to a measured value.

The uncertainties discussed above are referred to by many scientists as *experimental errors*, or simply *errors*. However, this terminology can be confusing, because in everyday usage an error is a mistake – something that is wrong. Even in the best scientific experiments, carried out with the utmost care by the most skilled scientists, there will be an uncertainty in a measured value; the negative connotations of the word 'error' seem to make the use of the term inappropriate here.

Another reason for avoiding the term *error* in this context is that quoting an error of $\pm x$ in a measurement implies that there exists a definite correct value. However, in many situations this is not the case. For example, you may be measuring a quantity that fluctuates with time, so that each measurement gives a different value. In cases like this, the scatter of the measured values indicates the variability of the quantity being measured.

For these reasons, we will talk about uncertainties rather than errors in this book. However, you should keep in mind that many authors, out of deference to convention, use the term error instead.

13.1 How do uncertainties arise?

It is useful to be aware of the various types of uncertainties that can occur in astronomical measurements, and to understand how and why uncertainties arise. This knowledge will enable you to recognize (and take steps to minimize) the uncertainties in the measurements that you make. While we cannot give a comprehensive list of all sources of uncertainty in astronomical observations, the categories outlined below should provide a framework for considering what the uncertainties might be in a particular situation.

Note that real errors (mistakes) are *not* included in this discussion. This is not because they never occur, but because they are impossible to predict or quantify. Everyone will misread a scale on occasions, transpose digits when writing down a number, or incorrectly apply a calibration factor to a measurement. However, you can generally avoid these errors by careful attention to the procedure that you are following and by always checking measurements and calculations.

(a) Uncertainties caused by lack of skill. This kind of uncertainty is one that almost falls into the 'mistake' category. The ability to start or stop a stopwatch to coincide with an event that is being timed is a skill that can be developed. Other uncertainties of this type can arise from not setting up an instrument correctly, or pointing at the wrong astronomical object! In general, such uncertainties become smaller as you gain more experience with astronomical work. Also, modern instruments are generally designed to minimize such uncertainties, for instance, when setting an exposure time for an observation, the detector will usually be pre-programmed to carry this out automatically without you having to start and stop it according to a stopwatch.

(b) Uncertainties caused by instrumental limitations. This kind of uncertainty is due to the nature of the equipment you are using and can really only be reduced by improving the quality of the equipment used. For instance, a CCD chip used as an astronomical detector may produce particularly 'noisy' signals. Its response to light of a certain brightness may vary randomly with time in a way that is not possible to compensate for using normal bias subtraction and flat-fielding techniques.

(c) Uncertainties caused by extraneous influences. A variety of unwanted effects can cause uncertainties in astronomical observations. Stray light from around the observatory may interfere with observations, changes of temperature can lead to uncertainties as the behaviour of a CCD depends on temperature. Again, you should attempt to eliminate such effects, or at least reduce them.

All of the uncertainties that have been mentioned so far have their origin in the measuring instruments or the measuring process. They could be present even if the quantity being measured had a precisely defined and unchanging value. But there are other rather different uncertainties – uncertainties that could be present even if it were possible to devise an ideal measuring instrument capable of infinite precision. These other uncertainties are caused by variations in what is actually being measured. Here are a couple of examples.

(d) Uncertainties caused by real variations in the quantity that is measured. In many astronomical observations, the quantity that is being measured is not 'fixed'. For example, in a set of observations to measure the apparent magnitude of a star, the star may (unknown to you) be a pulsating variable star whose magnitude varies with time. So if you repeated measurements of the star's apparent magnitude, making each measurement at a different time, then you might end up with a spread of values. As long as this spread was too large to attribute to instrumental limitations, then the spread would indicate the extent to which the star's apparent magnitude varies.

(e) Uncertainties caused by random fluctuations. This is really a subdivision of the previous category, and involves (random) variations with time of the quantity being measured. To take the example of measuring a star's magnitude again, a CCD simply records a voltage related to the number of photons arriving at a particular location on the detector within a given time interval. This number will vary slightly even if the star's apparent magnitude is constant. The light from a star arises from random atomic transitions in its surface layers, and the light then propagates across space for many years before entering the telescope. It is highly unlikely that *precisely* the same number of photons will arrive every second! Such fluctuations are commonly referred to as **noise**, and they are always present, superimposed on a signal that is of interest.

13.2 Random and systematic uncertainties

The uncertainties that were described in Chapter 13 Section 1 can be divided into two quite different types, those that are *random* and those that are said to be *systematic*.

A **random uncertainty** leads to measured values that are scattered in a random fashion over a limited range, as shown in Figure 13.1. The smaller the random uncertainty in the measurements, the smaller is the range over which they are scattered. Measurements for which the random uncertainty is small are described as **precise**.

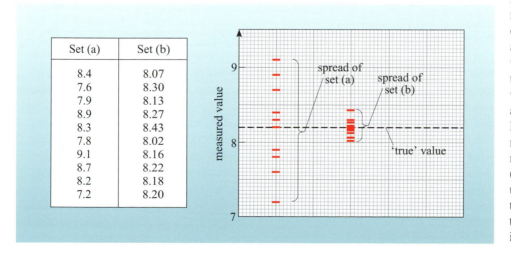

Set (a)	Set (b)
8.4	8.07
7.6	8.30
7.9	8.13
8.9	8.27
8.3	8.43
7.8	8.02
9.1	8.16
8.7	8.22
8.2	8.18
7.2	8.20

Figure 13.1 Two examples of random uncertainties. The two sets of measurements in the table, (a) and (b), are represented by the vertical positions of the dashes on the graph. The ten measured values for each set are scattered around the same 'true' value. However, the range over which the measurements are scattered is much larger for set (a) than for set (b). This indicates that the random uncertainty is greater for set (a) than for set (b), which means that the precision of the measurements is lower for set (a) than for set (b).

The best estimate that we can make for the value of the measured quantity is the mean, or average, of the measured values. As you might expect, if we make more measurements, then the mean value that we calculate is likely to be a better estimate of the quantity that we are measuring. We will make this statement quantitative later.

Systematic uncertainties have a different effect on measurements. A **systematic uncertainty** leads to measured values that are all displaced in a similar way from the true value, and this is illustrated in Figure 13.2. The two examples shown have the same random uncertainty – in both cases the spread, or scatter, of the values is the same. However, in both cases the measured values are systematically displaced from the true value. The values in set (b) are all larger than the true value, and the values in set (a) are all smaller. The difference between the mean value of a set of measurements and the true value is the systematic uncertainty. Measurements in which the systematic uncertainty is small are described as **accurate**. Therefore, to improve the accuracy of a measurement we need to reduce the systematic uncertainties.

The problem is that we don't generally know the true value, otherwise we would not need to make the measurement! So we need to estimate the possible sizes of systematic uncertainties by considering details of the apparatus and observational procedures. Alternatively, we need to devise a method of eliminating the systematic uncertainties.

Figure 13.2 The effect of systematic uncertainties. Two sets of measurements, (a) and (b), are represented by the vertical positions of the dashes on the graph. For set (a), the systematic uncertainty causes all of the measured values to be smaller than the true value. For set (b) the systematic uncertainty causes all of the values to be larger than the true value, but the size of the uncertainty is smaller than for set (a). The measurements in set (b) are therefore more accurate than those in set (a).

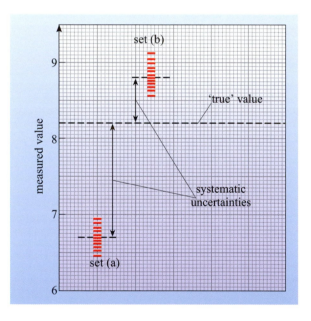

Now you might conclude that the goal of every astronomer is to ensure that their measurements are as *precise* as possible (i.e. smallest possible *random* uncertainties) and as *accurate* as possible (i.e. smallest possible *systematic* uncertainties). However, life is not that simple. Compromises always have to be made. How much time do you have to make the measurements? What measuring instruments are available, or what can you afford to buy? What is the purpose of the investigation? (After all, there is no point in trying to obtain a result that has a precision of ±0.01% if ±1% would be sufficient.) In addition, you may want to make a trade-off between accuracy and precision. Before these decisions can be made, you need to be able to estimate the accuracy and precision of measurements, or in other words, you need to be able to estimate the size of the systematic and random uncertainties.

One of the most difficult aspects of an astronomical observation can be the estimation of the uncertainties in the measurements. No two measurements are identical, so there are no definite rules about how you estimate the size of the uncertainties. However, it is very largely a matter of common sense. We will provide a few examples that illustrate how uncertainties can be quantified, and these should help you to decide how to approach estimating the uncertainties in work that you carry out.

In many experiments, more than one of the types of uncertainty discussed earlier will affect a measurement. Sometimes you will be able to estimate the size of each of the uncertainties individually and sometimes you will only be able to estimate their combined effect. But you should always keep in mind that it is only an *estimate* of the uncertainty that is required. By definition, an uncertainty cannot have an exact value, and it is generally sufficient to estimate the value of an uncertainty to one significant figure, or perhaps two. The uncertainty then allows you to decide how many significant figures to quote in a result. For example, if the result of a calculation is that $m_v = 12.345\,67$, and the uncertainty is calculated to be 0.043 21, then you should quote the result as $m_v = 12.35 \pm 0.04$. Here the uncertainty is rounded to one significant figure, and the best value is rounded to four significant figures, since then the uncertainty is affecting the last digit.

13.3 Estimating random uncertainties

We will start by considering random uncertainties, because they are often more straightforward to estimate than systematic uncertainties. Let's consider a specific experiment in which the magnitude of star is determined to be $m_v = 12.29$. What random uncertainty would we associate with such a measurement? Is it ±0.1, ±1, or perhaps ±0.02?

There are two distinct methods of estimating random uncertainties: one method involves repeating the measurement a number of times, and the other involves estimating the uncertainty from knowledge about the instruments and techniques used. It is good practice to use both methods and to check that they produce consistent estimates, but this will not always be feasible.

13.3.1 Estimating random uncertainties by repeating measurements

One way to estimate the size of random uncertainties in a measured value is by making a series of repeated measurements of the quantity. Random uncertainties lead to a scatter in measured values, and the uncertainty in the measurements can be deduced from the range over which the values are scattered. So for the magnitude measurement, introduced in the previous paragraph, we could make a series of, say, five measurements of the magnitude of the star. Suppose that the results were:

12.31, 12.26, 12.42, 12.25, 12.21

Let us assume that these measurements were all made with the same care and skill, and with the same detector. Then the best estimate we can make of the star's magnitude is the **mean** value of the five measurements, which is denoted by $\langle m_v \rangle$ and defined as:

$$\langle m_v \rangle = \frac{12.31 + 12.26 + 12.42 + 12.25 + 12.21}{5}$$

$$= 12.29$$

Now if there had been no random uncertainty associated with the measurement of the magnitude, then all of the five values would have been identical. The effect of the random uncertainty is to scatter the measurements around the true value, and the larger the random uncertainty, the greater will be the range over which the measurements will be scattered, i.e. the lower will be the precision of the measurements. The extent of the scatter therefore indicates:

- the size of the random uncertainty;
- how far from the true value a typical measurement is likely to be;
- conversely, how far from a measured value the true value might be;
- the precision of the measurement.

The five measurements are scattered between magnitudes of 12.21 and 12.42, which is a range of 0.21, or a spread of about ±0.1 around the mean. This spread of ±0.1 is one way of quantifying the random uncertainty of each measurement, and we would expect that additional measurements would lie roughly within this range.

However, this is a somewhat pessimistic estimate of the uncertainty in each measurement, because more of the measurements will lie in the centre of the range than lie at either extreme of the range. So, as a rough rule of thumb, we generally take the uncertainty in each measurement as about 2/3 of the spread of the values. In the example above, we would quote the uncertainty as $2/3 \times (\pm 0.1) = \pm 0.07$. This simple procedure is perfectly adequate in many cases. It is all that can be done, when, as above, a relatively small number of measurements is involved and it becomes quite reliable when more measurements are available.

So for the measurement of the star's apparent magnitude we could say that:

- the size of the random uncertainty is ±0.07 magnitudes;
- a typical measurement is likely to be within ±0.07 magnitudes of the true value;
- conversely, the 'true' value is likely to be within ±0.07 magnitudes of a measured value;
- the precision of the measurement is ±0.07 magnitudes.

Obviously if one measurement were very different from all the others, then using the spread to determine the uncertainty would give a misleadingly pessimistic value. Common sense suggests that the single deviant reading should be ignored when estimating the mean and the uncertainty, and suggests that a few more measurements should be taken. For example, if the last reading for the star's apparent magnitude were $m_V = 12.65$ rather than 12.21, then it would be wise to ignore it, or take further readings.

■ How else might you average a set of measurements in order to ignore the effects of a particularly discrepant value?

❏ You could calculate the *median* value of the set of measurements instead of the mean.

In the example above, the variability in the measured values of m_V could arise not only from uncertainties in the measuring or calibration process used, but also from real variations in the luminosity of the star. Even if it were possible to make exact measurements of the star's magnitude (i.e. we had a perfect measurement technique), there would be variability in the measured values if the flux from the star were varying – possibly due to real physical changes occurring in the star under investigation. And even if the flux from the star did not vary, then there could be variability in the measured values because of uncertainties in the measuring process, for example, uncertainties in timing the exposure or uncertainties in setting the aperture for the photometry. However, by repeating the measurements we get an overall measure of the random uncertainties, and it isn't necessary to make separate assessments of the random uncertainties due to these different factors. So, generally, you should not be satisfied with making a single measurement of a quantity, but should repeat the measurement several times.

Here is an important point that is worth emphasizing:

> The presence of a random uncertainty in a measurement can be detected – and its size estimated – by repeating the measurement a number of times.

13.3.2 Estimating random uncertainties from information about the instruments and techniques used

The second method of estimating random uncertainties only applies to uncertainties arising from the measuring process, and doesn't tell you anything about any variability in the quantity that is being measured. The method is particularly important when it is not possible to repeat measurements. It requires you to use your knowledge of, and experience with, the measuring instruments and the experimental techniques to estimate the likely random uncertainties.

Returning to the star brightness measurement, suppose that we had only made one measurement of the star's magnitude. We could then make estimates of some of the random uncertainties that might contribute to the overall uncertainty in the magnitude. For example, we might estimate on the basis of some other measurements, that there could be a random uncertainty of ±0.05 magnitudes due to the uncertain zero point on the magnitude calibration arising from the standard star observations.

13.4 The distribution of measurements with random uncertainties

When discussing just a few repeated measurements, we have characterized the scatter of the measured quantity by 2/3 of the spread of the values. As more measurements are made, they are likely to be scattered over approximately the same range, i.e. the spread will remain the same. However, the measured values are not evenly scattered throughout this range.

Let's consider a hypothetical set of observations that involve measuring the apparent magnitude of a star, and look at the distribution of a set of measurements. Figure 13.3a shows a histogram of the distribution of 10 measurements of apparent magnitude, where the height of each bar on the histogram represents the number of measurements within a certain 0.1 magnitude interval. So, for example, the tallest bar indicates that four of the measurements were between $m_V = 14.5$ and 14.6. The spread of the results is 0.6 magnitudes, or ±0.3 magnitudes, so we would say that the uncertainty in a single measurement was about two-thirds of this, i.e. ±0.2 magnitudes.

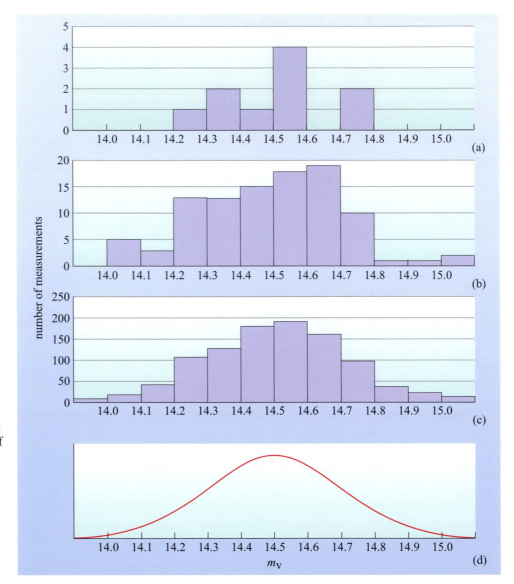

Figure 13.3 Histograms showing the distribution of measurements of the apparent magnitude of a star for (a) 10 measurements, (b) 100 measurements, (c) 1000 measurements. (d) The smooth curve that represents the distribution of a very large number of measurements separated into very small intervals of magnitude.

Figure 13.3b shows a distribution of 100 measurements. There is a smoother variation of the heights of the bars, but the spread of the distribution is similar to that for 10 measurements. Now imagine that we make 1000 measurements, as shown in Figure 13.3c. The distribution has taken on a characteristic bell shape. Continuing this process to the limit of a very large number of measurements, the envelope of the histogram bars might tend to become a smooth bell-shaped curve (Figure 13.3d).

13.4.1 The spread of the distribution

The distributions in Figure 13.3b–d all have extended wings on either side of the central peak. This means that using the overall spread of the measurements, or even 2/3 of the overall spread, as a measure of the random uncertainty may give a misleading estimate of how far a typical measurement lies from the mean, since the spread is calculated from only the maximum and minimum values. To avoid this problem, we need a measure of the random uncertainty that depends on the values of all of the measurements, not just the two most extreme. The quantity that is widely used for this purpose is the **standard deviation** of the measurements, and Box 13.1 shows an example of how the standard deviation is calculated.

BOX 13.1 CALCULATING THE STANDARD DEVIATION OF A SET OF MEASUREMENTS

The six steps below describe how the standard deviation is calculated. Table 13.1 is an example of this calculation, and the **BOLD** numbers on the table correspond to the numbered steps. As you read each step, you should refer to the corresponding part of the tabulated calculation.

1 Start with a set of measured values $x_1, x_2, x_3, \ldots x_n$. In the example in Table 13.1, $n = 8$.

2 (a) Add all of the values x_i and then (b) divide the sum by the number of values n to obtain the mean value $\langle x \rangle$ of the measurements:

$$\langle x \rangle = (x_1 + x_2 + x_3 + \ldots\ldots + x_n)/n = \Sigma x_i/n$$

3 The deviation d of a measurement x is defined as the difference between that measurement and the mean $\langle x \rangle$ of the set of measurements:

$$d = x - \langle x \rangle$$

You should now calculate the deviations d_i corresponding to each value x_i:

$$d_i = x_i - \langle x \rangle$$

4 Calculate the squares of each of the deviations, d_i^2.

5 (a) Add together all of the squares of the deviations d_i^2 and then (b) divide by the number of values n to obtain the mean square deviation:

$$\langle d_i^2 \rangle = \Sigma d_i^2/n$$

6 Take the square root of the mean square deviation. This is known as the standard deviation s_n:

$$s_n = \sqrt{\langle d_i^2 \rangle}$$

The standard deviation is generally quoted to one or two significant figures.

Table 13.1 An example of how the standard deviation of a set of measurements is calculated.

1 Measured value, x_i	3 Deviation, $d_i = x_i - \langle x \rangle$	4 Squared deviation, d_i^2
5.3	+0.01	0.0001
5.4	+0.11	0.0121
5.4	+0.11	0.0121
5.1	−0.19	0.0361
5.0	−0.29	0.0841
5.2	−0.09	0.0081
5.6	+0.31	0.0961
5.3	+0.01	0.0001
2a Sum of 8 values, $\Sigma x_i = 42.3$		**5a** Sum of squared deviations, $\Sigma d_i^2 = 0.2488$
2b Mean of 8 values, $\Sigma x_i/n = \langle x \rangle = 5.29$		**5b** Mean of squared deviations, $\langle d_i^2 \rangle = \Sigma d_i^2/n = 0.0311$
		6 Square root of mean squared deviations, $s_n = \sqrt{\langle d_i^2 \rangle} = 0.176,$ or $s_n = 0.2$ to 1 sig fig

The process of calculating the standard deviation described in Box 13.1 is an extended definition, but it is useful to have a more succinct definition.

The **standard deviation** s_n of a set of n measured values x_i is the square root of the mean of the squares of the deviations d_i of the measured values from their mean value $\langle x \rangle$.

$$s_n = \sqrt{\frac{\sum d_i^2}{n}} \qquad (13.1)$$

where the deviation d_i of the measured value x_i from the mean value $\langle x \rangle$ is

$$d_i = x_i - \langle x \rangle, \qquad (13.2)$$

and the mean value $\langle x \rangle$ of the measurements is

$$\langle x \rangle = \Sigma x_i / n \qquad (13.3)$$

The standard deviation is the most commonly used measure of the scatter of a set of measurements, and is used to quantify the likely random uncertainty in a single measurement.

■ The standard deviation is sometimes known as the root-mean-square (rms) deviation. Explain why this name is appropriate.

❏ The standard deviation is calculated as the square *root* of the *mean* of the *square* values of each of the individual deviations.

The standard deviation described above is also sometimes referred to as the **population standard deviation**. This is to distinguish it from a related quantity known as the **sample standard deviation** and defined by

$$s_{n-1} = \sqrt{\frac{\sum d_i^2}{n-1}} \qquad (13.4)$$

The sample standard deviation corrects the tendency to understate the uncertainty in measurements, especially when the sample size is small. For most practical purposes, the difference between the two definitions is insignificant though, and you should just stick to the definition given by Equations 13.1 to 13.3.

How do values for the standard deviation compare with what we would get by using the simple '2/3 spread' rule introduced earlier? Well, for the data in Table 13.1, the spread is 5.6 − 5.0 = 0.6, or ±0.3, so 2/3 of the spread is ±0.2. The standard deviation calculated in Table 13.1 is 0.176, or 0.2 to one significant figure. This illustrates why the simple rule is adequate for many situations where we only need a rough estimate of the uncertainty of a measurement.

13.4.2 The shape of the distribution

In real situations, as opposed to hypothetical ones, astronomers rarely make sufficient measurements to obtain a smooth distribution. Even if they did, the shape of the distribution curve might depend markedly on the particular type of measurements that were being made. A useful model to describe how often you will count a certain number of occurrences of an event (like the detection of a photon) in a certain time interval is the **Poisson distribution**. The Poisson distribution can be used to describe a large variety of phenomena that are relevant to astronomy, however it is only applicable when each event counted is independent of all the other events. The Poisson distribution is not symmetric about its mean value, but as the number of events increases, it does become symmetrical, and approaches the shape of a standard mathematical form known as the **normal distribution**, which is also known as the **Gaussian distribution**. This is the distribution already shown in Figure 13.3d. Notice also that plotting on a magnitude scale, which is logarithmic, distorts a distribution to some extent.

Figure 13.4 shows how the standard deviation of a Gaussian distribution curve is related to the spread of the curve. It is clear that a substantial fraction of measurements deviates from the mean value by more than the standard deviation s_n. For a particular range of the measured variable, the area under the distribution curve represents the fraction of measurements that lie within that range. For a Gaussian distribution, 68% of measurements lie within one standard deviation, i.e. within $\pm s_n$, of the mean value. Therefore, 32% of measurements are expected to differ from the mean by more than the standard deviation s_n. Note that the distribution curve falls off rapidly as the measurements deviate further from the mean. Table 13.2 shows the percentage of measurements falling within specified ranges centred on the mean value.

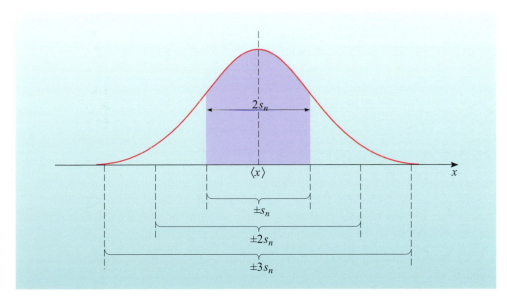

Figure 13.4 The standard deviation s_n characterizes the width of the Gaussian distribution. The shaded area under this Gaussian distribution curve represents the measurements that lie within $\pm s_n$ of the mean. This area is 68% of the total area under the curve, indicating that 68% of measurements are expected to fall within this range, hence the 'rule of two-thirds' used in Section 13.3.1.

Table 13.2 The percentages of measurements within, and outside, various ranges of values centred on the mean for a Gaussian distribution.

Range centred on mean value	$\pm s_n$	$\pm 2s_n$	$\pm 3s_n$	$\pm 4s_n$
Measurements within range	68%	95%	99.7%	99.994%
Measurements outside range	32%	5%	0.3%	0.006%

It is important to bear these percentages in mind when the standard deviation is used to indicate the uncertainty in a measurement. The statement that $m_V = 14.5 \pm 0.2$ does not mean that *all* measurements of the quantity m_V will lie within the range from 14.3 to 14.7. If the standard deviation is 0.2 magnitudes, then on average only 68% of measurements will lie within that range and 32% will lie outside. Therefore, roughly two-thirds of all measurements will lie within one standard deviation (i.e. within $\pm s_n$) of the mean, and one-third of all measurements will lie more than one standard deviation from the mean.

The standard deviation is a measure of the precision of measurements. The greater the precision, the smaller will be the scatter and therefore the smaller will be the standard deviation. This means that, with data of greater precision, the Gaussian distribution curve will have a much narrower peak around the mean value. This is illustrated by the three distribution curves shown in Figure 13.5.

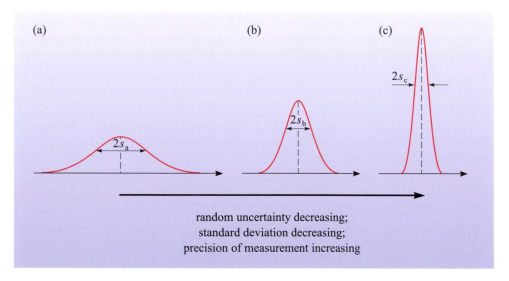

Figure 13.5 Gaussian distribution curves representing three large sets of measurements. The broadest and least highly peaked of the three curves (a) has the largest standard deviation, and corresponds to measurements with the largest random uncertainties and with the lowest precision.

random uncertainty decreasing;
standard deviation decreasing;
precision of measurement increasing

13.5 Uncertainties when counting randomly occurring events

An important type of random uncertainty arises when investigating processes that involve counting events that fluctuate randomly, such as the number of photons from a star arriving on a CCD detector. In this section you will see that this type of uncertainty can easily be calculated using a simple expression.

Let's consider just such an experiment in which a CCD is used to determine the number of photons arriving each minute from a particular star. Suppose that, after determining the ratio between output voltage and photon count in a particular measurement, it is calculated that the CCD recorded 9986 photons in one minute in the patch of the field containing the image of the star. How many photons would you record if you repeated the measurement for another one-minute period?

Since each individual arriving photon is an event subject to random fluctuations, the number of photons arriving in a period of one minute will vary. However, measurements of the number of photons arriving each minute will be clustered around a well defined mean value $\langle n \rangle$. In fact, if you repeated the measurement 1000 times you might find that the number of photons arriving per minute was distributed as shown by the histogram in Figure 13.6a.

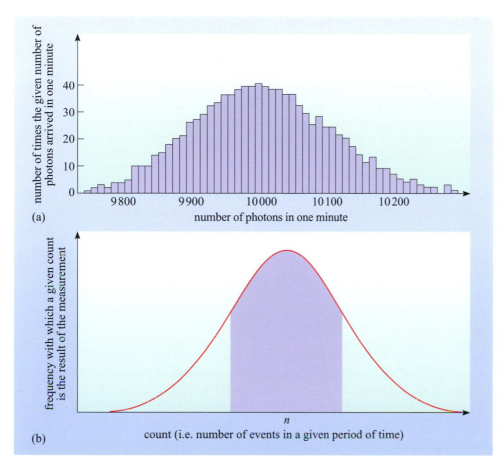

Figure 13.6 (a) A histogram showing how often particular numbers of photons were recorded from a certain star in 1000 counting periods of 1 minute. (b) As long as the mean number of events per counting period is large, then the distribution of the results from a very large number of measurements approximates to a Gaussian distribution curve. The mean $\langle n \rangle$ corresponds to the centre of the peak and 68% of all the measurements are within $\pm \sqrt{n}$ of this mean.

The mean $\langle n \rangle$ of the 1000 measurements corresponds to the position of the centre of the peak of this symmetrical distribution, i.e. 10 000 photons per minute. If you were to calculate the value of the standard deviation from the results in the histogram you would get a value of 100 photons per minute. As you saw earlier, the standard deviation is used as a measure of the random uncertainty in a single measurement, and so we can say that the uncertainty in this case is ±100. It is *not* a coincidence that this uncertainty of ±100 is the same as the square root of the mean number of photons, 10 000.

Now imagine that you make many successive measurements of the number of randomly occurring events (such as photons arriving from a star) in a given period. As long as the mean number of events per counting period is large, then as the number of measurements increases, the envelope of the histogram of the results will tend to a Gaussian distribution curve, as shown in Figure 13.6b. The centre of the distribution indicates the mean number $\langle n \rangle$ of events in the given period. More interestingly in the present context, the standard deviation of the measurements is $\sqrt{\langle n \rangle}$, and so, as was shown in Figure 13.4, 68% of the measurements lie within $\pm \sqrt{\langle n \rangle}$ of the mean. For example, if the mean is 10 000 photons, then 68% of the measurements will produce a result between 9900 and 10 100 photons. Alternatively, if the mean is 900 photons, 68% of the results will lie between 870 and 930. The standard deviation $\sqrt{\langle n \rangle}$ defines the width of the distribution, and it is therefore straightforward to identify the uncertainty associated with a single counting measurement.

If the number of *randomly fluctuating events* counted in a given period is n, the uncertainty in this number is given by:

$$\text{uncertainty} = \sqrt{n} \qquad (13.5)$$

This uncertainty is a measure of the likely difference between the value n that would be counted in any *single* measurement and the mean value of *many* measurements of n, namely $\langle n \rangle$, that would be found from a long series of repeated measurements.

Now Equation 13.5 indicates that the uncertainty increases as the number of events counted increases. With the example we have used, about 10 000 events were counted in a one-minute interval, and the uncertainty is about ±100. In a six-second period, only about 1000 events would be counted, with an uncertainty of ±32. Increasing the counting interval to 10 minutes would give about $(100\,000 \pm 320)$ events, and 100 minutes would give about $(1\,000\,000 \pm 1000)$ events.

■ Do these numbers suggest that it is better to count for a shorter period than for a longer period?

❏ No. It is true that the uncertainty in the number of events gets smaller as the counting period gets shorter (and the number of events gets smaller). However, the uncertainty $\sqrt{n}$ becomes a much larger *fraction* of the number of events n as the counting period gets shorter. By counting for a longer time we can reduce the uncertainty $\sqrt{n}$ as a fraction of the number of counts n.

It is important to note that *increasing* the number of events *reduces* the fractional uncertainty:

$$\text{fractional uncertainty} = \frac{\text{uncertainty}}{\text{measured value}} = \frac{\sqrt{n}}{n} = \frac{1}{\sqrt{n}} \qquad (13.6)$$

This reduction in the fractional uncertainty as the counting period (and hence the number of events) increases is demonstrated by the data in Table 13.3.

Table 13.3 Uncertainties associated with counting random events for different intervals, with the same mean event rate (100 events per minute) in each case. Although the uncertainty *increases* as $\sqrt{n}$ (third row) as the counting interval *increases*, the fractional uncertainty $\sqrt{n}\,/\,n$ (fourth row) *decreases*. All values are quoted to one significant figure.

Counting interval t/minute	0.1	1	10	100
Typical number of events, n	10	100	1000	10 000
Uncertainty in number of events, $\sqrt{n}$	3	10	30	100
Fractional uncertainty in number of events, $\sqrt{n}\,/\,n = 1/\sqrt{n}$	0.3	0.1	0.03	0.01
Uncertainty in number of events per minute $= \sqrt{n}\,/\,t$ /minute^{-1}	30	10	3	1

Another way to appreciate the improvement that results from counting for a longer interval is to compare the values of the uncertainty in the number of events per minute, which are displayed in row 5 of Table 13.3. These were calculated by dividing the uncertainties in row 3 by the corresponding interval in row 1: they show a similar improvement with increasing time (and number of events) to that shown by the fractional uncertainties.

Unfortunately, the rate at which the fractional uncertainties improve with increasing n is frustratingly slow. For example, to reduce the fractional uncertainty by a factor of 10, the number of events must be increased by a factor of 100. Of course, that requires counting for an interval that is 100 times longer. So one of the skills every astronomer needs to develop is the ability to decide how to balance the time invested in an observation against the precision and accuracy of the result.

13.6 Estimating systematic uncertainties

Systematic uncertainties and random uncertainties are often both present in the same measurement, and their effects were illustrated in Figure 13.1 and Figure 13.2. The spread of repeated measurements allows us to estimate the size of the random uncertainties, and averaging the measurements tends to cancel out the effects of such uncertainties. Unfortunately, systematic uncertainties are usually much more difficult to estimate. Repeated readings do not show up the presence of systematic uncertainties, and no amount of averaging will reduce their effects.

Systematic uncertainties often arise from the measuring instrument used. For example, a metre rule may in fact be 1.005 m long, so that all measurements made with it are systematically 0.5% too short. Systematic uncertainties like this can often be discovered and estimated by calibrating the measuring instrument against a more accurate and reliable instrument. If this can be done, the measured results can be corrected and so the effects of this type of systematic uncertainty can be reduced, possibly to a level that is negligible compared with other uncertainties.

Another common type of systematic uncertainty is a 'zero' uncertainty. The dark current present on a CCD image is an example of this type of effect – a signal may be recorded on the CCD even when no light falls upon it. However, once noted and recorded, this type of systematic uncertainty is straightforward to eliminate by subtracting the zero error from the measured values.

It is important here to distinguish between systematic uncertainties that you can measure and allow for – and which, therefore, will *not* contribute to the uncertainty in the final result – and systematic uncertainties for which you can only say that they are 'likely to be $\pm x$'. For example, in the case of measuring a star's apparent magnitude earlier, the two different kinds of systematic uncertainty could occur when timing the exposure of a CCD image. If you were doing this manually using a stopwatch, then perhaps the watch might run slow. Comparing it to a more accurate clock might show that it lost 10 s in a half-hour exposure, which means that exposure times need to be scaled up by a factor of 1800/1790 to get the actual exposure time. If this calibration correction is made, then the slow running of the stopwatch would not contribute to the uncertainties in the observation. The calibration procedure eliminates this systematic uncertainty. However, you might have a tendency to start or stop the stopwatch too early or too late each time. You wouldn't *know* that such a systematic uncertainty was present, but it is certainly possible. A reasonable estimate for the possible size of such an uncertainty is ±0.2 s (a typical human response time), since anything longer would probably be detected. This uncertainty of ±0.2 s in each exposure (let us suppose of 10 minutes), or ±1 part in 3000, would lead to an uncertainty of ±1 part in 3000 in the magnitude of the star that is measured.

So the difference between these two types of systematic uncertainty is that we know that one is definitely present, and we can measure and correct for its effect, whereas the other may or may not be present, and we can only make an educated guess at its possible size. Essentially, once we have identified, measured and corrected for the first type of uncertainty, it ceases to be a source of uncertainty in the final result.

So your aim as an astronomer is to look critically at your observations – at the instruments, the methods of measurement, the techniques that you use – and to identify and quantify the uncertainties that may be present. You may need to calibrate instruments, or make measurements with different instruments, or use alternative techniques, in order to do this. You can then decide whether or not you need to attempt to reduce the systematic uncertainties.

13.7 Combining uncertain quantities

In most astronomical observations, more than one source of uncertainty will be present. Several random and systematic uncertainties may contribute to the uncertainty in measurements of a single quantity. In addition, measurements of a number of different quantities (each of which has an uncertainty associated with it) may have to be combined to calculate the required result. Therefore, it is important to know how these different uncertainties are combined to determine the overall uncertainty in the final result.

It is not our aim here to get involved in statistical theories, so the rules for combining uncertainties will be presented without proof. The major objective is that you should be able to choose and apply the appropriate methods for combining uncertainties in the work you are engaged in. You should however be aware that all the rules given depend on the assumption that the sources of error are random and independent, which is not always true!

13.7.1 The uncertainty in a mean value

The standard deviation s_n of a set of measurements tells us about how widely scattered the measurements are – it indicates how far the individual measurements are likely to be from the mean value. We usually take the mean value of the measurements as our best estimate of the true value, and so what we really need to know is how far the mean value is likely to be from the true value. In other words, we want to know the uncertainty in the mean value.

We'll consider again the hypothetical set of observations of the flux from a star, which was introduced earlier, and we will assume that the Gaussian distribution in Figure 13.7a represents the distribution of a very large number of measurements of this star's flux. (It is the *flux* that is measured. Taking the logarithm of it to get a *magnitude* means that the Gaussian is distorted.) Suppose that we make five measurements of the flux from the star, and calculate the mean value of these five measurements. We then repeat this process nine times, so that we end up with ten values of the mean, each of which is based on five different measurements. Figure 13.7b shows what these measurements and their mean values might look like.

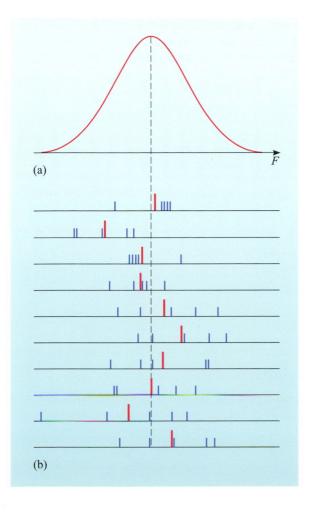

(a)

(b)

Figure 13.7 (a) A Gaussian distribution curve for measurements of the flux from a particular star. (b) The blue bars marked on each of the ten scales indicate five measurements of the flux from the star, and each red bar indicates the mean value of a set of five values.

■ Is the spread of the ten red bars that indicate the mean values in Figure 13.7b greater than, smaller than, or the same as, the spread of the individual measurements of the star's flux? Explain the reason for your answer.

❏ The spread of the means from the ten sets of measurements, that is, the spread of the red bars in Figure 13.7b, is *smaller* than the spread of individual measurements, which is represented by the width of the distribution curve shown in Figure 13.7a. This is because a set of five measurements will almost certainly include some that are greater than the true mean $\langle F \rangle$ and some that are smaller. Therefore, when the mean of five measurements is calculated, it will be closer to the true mean $\langle F \rangle$ than most of the five individual measurements are.

■ Suppose that you repeated the process described above, but with sets of 20 measurements, rather than sets of five measurements. Would you expect the spread of the means from the ten sets of 20 measurements to be greater than, smaller than, or the same as, the spread of the means from the ten sets of five measurements?

❏ The means from the sets of 20 measurements will have a *smaller* spread than the means from the sets of five measurements. The larger the number of measurements in a set, the smaller the statistical fluctuations in their mean value, and the closer the mean will lie to the mean of a very large number of measurements.

This is illustrated in Figure 13.8, which shows the distribution curve from Figure 13.7a, together with the ten values of the means from the sets of five measurements (Figure 13.8b) and ten values of the means from sets of 20 measurements (Figure 13.8c). The means from the five-measurement sets are less widely scattered than the distribution of individual measurements, and the means from the 20-measurement sets are even less scattered.

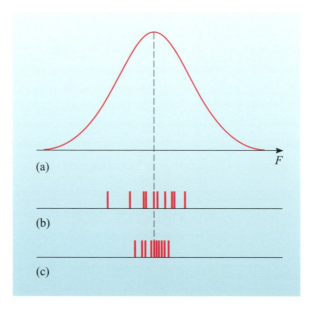

Figure 13.8 (a) The Gaussian distribution curve for measurements of the flux from a particular star. (b) Values of the means from ten sets of five measurements each; these means are less spread out than the individual measurements. (c) Values of the means from ten sets of twenty measurements each; these means are less spread out than the means of five measurements shown in (b).

This is an important result, with far-reaching implications.

> The uncertainty in a mean value *decreases* as the number of measurements used to calculate the mean *increases*.

In other words, you can reduce the uncertainties in an experiment by increasing the number of measurements that you make.

Let's now make this statement about the uncertainty of a mean value quantitative.

> The uncertainty σ_m in a mean value that is derived from n measurements that have a standard deviation s_n is
>
> $$\sigma_m = \frac{s_n}{\sqrt{n}} \qquad (13.7)$$
>
> We will refer to σ_m as the **uncertainty in the mean**. It is usually referred to as the *standard error in the mean*, but as we explained earlier we prefer to avoid the term error in this context.

As noted earlier, there are two slightly different definitions of standard deviation. The population standard deviation has $\sqrt{n}$ on the bottom line, while the sample standard deviation has $\sqrt{n-1}$. Similarly, the uncertainty in the mean can also be defined as

$$\sigma_m = \frac{s_{n-1}}{\sqrt{n-1}} \tag{13.8}$$

This latter formula gives a more realistic result when the number of measurements is small. However, both formulae give similar results with large n, and it is probably best to stick to Equation 13.7.

It is important to note the effect of the square root that appears in Equation 13.7. If we take five measurements, then $\sqrt{n} = 2.2$, so $\sigma_m = 0.45s_n$. If we take ten times as many measurements, i.e. 50 measurements, then $\sqrt{n} = 7.1$, so $\sigma_m = 0.14s_n$. So for the increased effort of taking ten times more measurements, we only reduce the uncertainty in the mean by a factor $\sqrt{50}/\sqrt{5}$, which is $\sqrt{10}$ or ~ 3. It is also important to realize that breaking up a series of measurements into sets has *no effect* on the overall uncertainty – take all 50 together, or ten sets of five, or five sets of ten – the uncertainty in the mean is the same.

The significance of the uncertainty in the mean can be brought out in the following way. Suppose we make many sets of n measurements, and evaluate the mean for each set. Then the width of the distribution of the means will be characterized by the uncertainty in the mean σ_m rather than by the standard deviation s_n. So whereas the standard deviation tells us about the scatter of individual measurements, the uncertainty in the mean of n measurements tells us about the scatter of the mean values that are each derived from n measurements.

If we assume that the mean values have a Gaussian distribution, then we can be more explicit about the meaning of the uncertainty in the mean σ_m. Suppose that the mean value is 10.4 ± 0.3, where $\sigma_m = 0.3$ is the uncertainty in the mean. This does *not* indicate that the true value necessarily lies within ± 0.3 of the mean value of 10.4; the uncertainty in the mean is *not* like an engineering tolerance. If the mean values have a Gaussian distribution, then there is a 68% probability that the calculated mean value lies within $\pm\sigma_m = \pm0.3$ of the true mean value, a 95% probability that it lies within $\pm2\sigma_m = \pm0.6$ of the true mean value, and a 99.7% probability that it lies within $\pm3\sigma_m = \pm0.9$ of the true mean value. Conversely, we can say that there is a 68% probability that the true value lies within $\pm\sigma_m$ of the calculated mean value, a 95% probability that the true value lies within $\pm2\sigma_m$ of the calculated mean value and a 99.7% probability that the true value lies within $\pm3\sigma_m$ of the calculated mean value.

13.7.2 Combining uncertainties in a single quantity

Let's return to the example measuring the magnitude of a star, and consider how we would combine a systematic uncertainty of ±0.05 magnitudes arising from the calibration process and a random uncertainty of ±0.04 magnitudes in the measured brightness. The obvious answer might seem to be to add them directly to get a total uncertainty of $0.05 + 0.04 = 0.09$. But this really gives an unduly pessimistic assessment of the uncertainty. After all, the random uncertainty and the systematic uncertainty are assumed to be entirely *independent*, so it is highly unlikely (though possible, of course) that, for any given measurement, these two uncertainties will both be at their maximum positive values, or at their maximum negative values. There will generally be a partial cancellation of the two uncertainties. So the rule that we use for combining two *independent* uncertainties in the *same* quantity is

$$\delta X = \sqrt{\delta x_1^2 + \delta x_2^2} \tag{13.9}$$

where δX is the overall uncertainty, and δx_1 and δx_2 are the individual uncertainties that are to be combined. Thus, in the example of the star's apparent magnitude,

$$\begin{aligned}
\delta X &= \sqrt{0.05^2 + 0.04^2} \\
&= \sqrt{0.0025 + 0.0016} \\
&= \sqrt{0.0041} \\
&= 0.06 \text{ magnitudes}
\end{aligned}$$

This uncertainty is larger than either of the contributing uncertainties, but smaller than their sum.

If more than two uncertainties are involved, then the method follows the same principle. Suppose that we think that there is an additional systematic uncertainty of ± 0.03 magnitudes in measuring the brightness of the star. Then the overall uncertainty is

$$\begin{aligned}
\delta X &= \sqrt{0.05^2 + 0.04^2 + 0.03^2} \\
&= 0.07 \text{ magnitudes}
\end{aligned}$$

Again, this is larger than the individual uncertainties, but smaller than their sum.

To summarize:

Independent uncertainties δx_1, δx_2, δx_3, ... in a measured quantity will give rise to an overall uncertainty δX given by

$$\delta X = \sqrt{\delta x_1^2 + \delta x_2^2 + \delta x_3^2 + \ldots} \tag{13.10}$$

When uncertainties are not independent, they are much more difficult to deal with, unless the form of the dependence is known precisely. Their dependence might mean, for example, that a large positive uncertainty from one source was always associated with a large negative uncertainty from another source, thus leading to cancellation, and an overall uncertainty that was much smaller than the individual uncertainties. Alternatively, a positive uncertainty from one source may always be associated with a positive uncertainty from another source, so that the overall uncertainty is really more like the sum of the individual uncertainties. No simple rules can be given for dealing with dependent uncertainties, and each case must be analysed individually.

13.7.3 Combining uncertainties in sums, differences, products, ratios and powers

In the last section, we were concerned with combining uncertainties in a *single* measured quantity to find the total uncertainty in that quantity. However, the aim of many investigations is to evaluate something that depends on *several* measured quantities, *each* of which has its own uncertainty.

Table 13.4 gives a set of rules, from which you can select the appropriate one to apply in a given situation.

Table 13.4 Rules for combining uncertainties. The first column lists various relationships between a quantity X and measured quantities A, B, which have uncertainties δA, δB. The second column indicates how the uncertainty δX in X is related to the uncertainties δA, δB. The symbols j, k and n represent constants (i.e. they have no uncertainty associated with them).

Dependence of X on A and B	Expression used to calculate δX	
$X = kA$	$\delta X = k\,\delta A$, or equivalently $\dfrac{\delta X}{X} = \dfrac{\delta A}{A}$	(13.11)
$X = kA + jB$ $X = kA - jB$	$\delta X = \sqrt{(k\,\delta A)^2 + (j\,\delta B)^2}$	(13.12)
$X = kA \times jB$ $X = kA / jB$	$\dfrac{\delta X}{X} = \sqrt{\left(\dfrac{\delta A}{A}\right)^2 + \left(\dfrac{\delta B}{B}\right)^2}$	(13.13)
$X = kA^n$	$\dfrac{\delta X}{X} = n\dfrac{\delta A}{A}$	(13.14)

There are several important points to note about the expressions for the uncertainties in Table 13.4.

- Only + signs appear under the square roots in the expressions for δX and $\delta X/X$, irrespective of whether the function is a sum, difference, product or ratio.

- The constants j and k have *no* effect on the *fractional* errors in Equations 13.11, 13.13 and 13.14.

- Note also that *percentage* uncertainties can be used in place of *fractional* uncertainties in the equations in Table 13.4.

13.7.4 Uncertainties in astronomical magnitudes

As well as combining uncertainties in quantities that appear as sums, differences, products, ratios or powers, in astronomy you will also need to evaluate uncertainties in astronomical magnitudes based on uncertainties in the measured flux, i.e. the photon count recorded on the CCD. Since the relationship between astronomical magnitude and flux involves a logarithmic function, the procedure is a little more complicated and deserves a careful explanation.

Earlier you saw that Equation 6.1 gives the general relationship between the magnitudes and fluxes of two astronomical objects. We can simplify this expression to write the astronomical magnitude m of a single object in terms of its flux F as

$$m = -2.5 \log_{10}(F) + K \tag{13.15}$$

where K is a constant term that is determined by the observational set-up you are using. Now, the flux is simply the number of photons arriving on the CCD. For simplicity, let's assume it's the number of photons measured within a certain aperture around a star, after background subtraction and all the usual CCD calibration steps have been performed. What then is the uncertainty in the value of the astronomical magnitude?

For any mathematical function $X = f(A)$ the uncertainty in X, indicated by δX, is related to the uncertainty in the quantity A, indicated by δA, by the relationship:

$$\delta X = f(A + \delta A) - f(A) \tag{13.16}$$

In other words, the uncertainty in X is equal to the value of the function evaluated at $(A + \delta A)$ minus the value of the function evaluated at A.

In the specific case of astronomical magnitudes we may work this out as follows:

$$\delta m = [-2.5 \log_{10} (F + \delta F) + K] - [-2.5 \log_{10} (F) + K]$$

$$= -2.5 \log_{10} (F + \delta F) + 2.5 \log_{10} (F)$$

$$= -2.5 [\log_{10} (F + \delta F) - \log_{10} (F)]$$

$$= -2.5 \log_{10} \left(\frac{F + \delta F}{F} \right)$$

Note that we can neglect the minus sign here, as we're only interested in the size of the uncertainty in m, so the general equation is:

$$\delta m = 2.5 \log_{10} \left(1 + \frac{\delta F}{F} \right) \tag{13.17}$$

Now, in the case where we're dealing with a *single* flux measurement whose only source of uncertainty is *random* counting statistics, $\delta F = \sqrt{F}$ for large fluxes, so this becomes

$$\delta m = 2.5 \log_{10} \left(1 + \frac{\sqrt{F}}{F} \right) \tag{13.18}$$

$$= 2.5 \log_{10} \left(1 + \frac{1}{\sqrt{F}} \right)$$

So, if the CCD records a flux of 10 000 photons from a particular star, the uncertainty in the flux due to counting statistics is ±100 photons (i.e. 1% of the value). The uncertainty in the star's instrumental magnitude is therefore $2.5 \log_{10}(1 + 1/100) \sim 0.01$. This is a useful rule of thumb:

> An uncertainty of 1% in the measured flux leads to an uncertainty of 0.01 magnitudes.

■ What flux would you require in order to obtain an uncertainty in the magnitude of ±0.05?

❑ We have

$$0.05 = 2.5 \log_{10} \left(1 + \frac{1}{\sqrt{F}} \right)$$

so, $10^{0.05/2.5} = 1 + \frac{1}{\sqrt{F}}$

$$10^{0.05/2.5} - 1 = \frac{1}{\sqrt{F}}$$

$$F = \left(\frac{1}{10^{0.05/2.5} - 1} \right)^2$$

$$F = 450$$

A list of uncertainties in magnitudes corresponding to a given flux of photon counts is given in Table 13.5. Note, these are only the *random* uncertainties due to counting statistics in the flux from the object itself.

Table 13.5 Uncertainties in magnitudes corresponding to a given flux of photon counts.

Flux / (photon counts)	Uncertainty in flux/ photon counts	Uncertainty in magnitude
100 000	300	0.003
10 000	100	0.01
1000	30	0.03
100	10	0.1

Although we have explained this in some detail, most software packages that perform aperture photometry on CCD images will carry out this calculation automatically for you. At the click of a button they will measure the background subtracted flux inside an aperture and display the result in astronomical magnitudes with its associated uncertainty value due to counting statistics. Nonetheless, it is important that you appreciate what is going on 'behind the scenes' to produce these numbers.

13.7.5 Signal-to-noise calculations in astronomical photometry

The previous section simply considered the uncertainty in the flux that arises as a result of the uncertainty in the number of photons detected from the star itself. However, in reality there will be other sources of uncertainty, or noise, when determining the brightness of an astronomical object. As explained in Chapter 5, these are:

- **sky noise**: there will be an uncertainty in the number of photons in the sky aperture that is subtracted from the target aperture.

- **thermal noise**: this is the dark current referred to in Chapter 5 Section 2 and is due to thermal motions in the CCD itself knocking some electrons free and causing them to be recorded as an additional count rate.

- **readout noise**: this is the random noise added by the CCD electronics as the signal from each pixel is read out. The value of the readout noise is generally a known quantity for any particular CCD and is expressed as a number of electrons.

Since the sky noise and the thermal noise both depend on the area of the aperture and pixel size and both increase as the exposure time increases, they can be combined into a single measure as the background noise per pixel.

Let us suppose the total number of photons in the target aperture is N_{star} and the number of photons in the background aperture is N_{back} per pixel, where the background aperture encompasses np pixels. Further suppose that the readout noise is R electrons per pixel. The uncertainty due to the number of photons in the target aperture is $\sqrt{N_{star}}$, and the uncertainty due to the number of photons in the background aperture is $\sqrt{N_{back} \times np}$. Then, following the rules in Table 13.4, the total uncertainty or noise in the measurement of the brightness of a target is

$$\text{noise} = \sqrt{N_{star} + (N_{back} \times np) + (R^2 \times np)} \tag{13.19}$$

The signal-to-noise ratio is then simply

$$\text{signal-to-noise} = N_{\text{star}}/\text{noise} \qquad (13.20)$$

Having calculated the signal-to-noise ratio, the uncertainty in the magnitude value may be determined using Equation 13.17. The following question gives an example of this type of calculation.

■ Suppose a target aperture contains 36 000 photons, and the background aperture on the CCD image contains 1100 photons per pixel and encompasses 24 pixels. The readout noise is 6 electrons per pixel.

(a) What is the signal-to-noise ratio of the measured flux of the target?

(b) What is the uncertainty in magnitudes of this measurement?

❑ (a) The total noise in the measurement is

$$\text{noise} = \sqrt{36\,000 + (1100 \times 24) + (6^2 \times 24)} = 252$$

So the signal-to-noise ratio of the measurement is 36 000/252 ~ 140.

The uncertainty in magnitudes is given by

$$\delta m = 2.5 \log_{10}\left(1 + \frac{\delta F}{F}\right) = 2.5 \log_{10}\left(1 + \frac{252}{36\,000}\right) = 0.0075$$

or better than one-hundredth of a magnitude.

In this example, the target is quite bright, so the readout noise is negligible, but the noise from the target itself and the background noise are comparable in size.

13.8 Some common sense about uncertainties

The examples in the previous subsection illustrate a few general points that are worth highlighting. First, uncertainties, by their very nature, cannot be precisely quantified. So a statement like $m_V = (8.732 \pm 0.312)$ is rather silly, and this result should be quoted as $m_V = (8.7 \pm 0.3)$. As a general rule:

Uncertainties should usually be quoted to one significant figure; two significant figures are sometimes justified, particularly if the first figure is a 1.

You should bear this in mind when trying to assess the size of uncertainties and when doing calculations involving uncertainties.

Secondly, you can safely neglect small uncertainties. The total uncertainty in a result may be a combination of several contributing uncertainties, and these contributing uncertainties may have widely varying sizes. But, because the uncertainties (or fractional uncertainties) are combined as the sum of the squares, as a general rule:

When calculating uncertainties in *sums and differences*, ignore any uncertainties that are less than 1/3 of the largest uncertainty.

When calculating uncertainties in *products and ratios*, ignore any *fractional* uncertainties that are less than 1/3 of the largest *fractional* uncertainty.

Thirdly, concentrate your efforts on reducing the dominant uncertainties. As we have shown, the largest uncertainties will dominate the uncertainty in the final result, and small uncertainties can often be neglected. Therefore, it is very important not to waste a lot of time reducing small uncertainties when much larger uncertainties are present.

Find out as early as possible in an investigation what the dominant uncertainties are, and then concentrate your time and effort on reducing them.

Finally, take particular care when differences and powers are involved. Suppose that you measure two angles, $\theta_1 = (73 \pm 3)$ degrees and $\theta_2 = (65 \pm 3)$ degrees, and you then calculate the difference, i.e. $\theta = \theta_1 - \theta_2 = 8$ degrees. The uncertainty is

$$\begin{aligned}
\delta\theta &= \sqrt{(\delta\theta_1)^2 + (\delta\theta_2)^2} \\
&= \sqrt{3^2 + 3^2} \\
&= \sqrt{18} \text{ degrees} \\
&= 4°
\end{aligned}$$

So $\theta = (8 \pm 4)$ degrees. This is a 50% uncertainty compared with only about 4% in the individual measurements!

To take another example, suppose you measure an edge of a cube as $l = (6.0 \pm 0.5)$ mm, and then calculate the volume: $V = l^3 = 216$ mm^3. The uncertainty is given by

$$\frac{\delta V}{V} = \frac{3\,\delta l}{l} = \frac{3 \times 0.5}{6}$$

Because the volume is the third power of the length, the fractional uncertainty in the volume is three times greater than the fractional uncertainty in the length measurement. As a general rule:

If a calculation involves taking the difference of two nearly equal measured quantities, or taking the power of a measured quantity, then pay particular attention to reducing the uncertainties in those quantities.

13.9 Summary of Chapter 13 and Questions

- *Random* uncertainties affect the *precision* of a measurement; *systematic* uncertainties affect the *accuracy* of a measurement.

- Random uncertainties may be estimated by repeating measurements. The best estimate of the measurement is the *mean* value: $\langle x \rangle = \Sigma x_i / n$ and the size of the random uncertainty in any individual measurement is about 2/3 of the spread of the measurements.

- The *standard deviation* s_n of a set of measured values x_i is the square root of the mean of the squares of the deviations of the measured values from their mean value:

$$s_n = \sqrt{\frac{\sum(x_i - \langle x \rangle)^2}{n}}$$

- In the limit of many measurements, the typical distribution of a set of measurements will follow a *Gaussian* (normal) distribution. 68% of the measurements will lie within ±1 standard deviation of the mean value.

- When counting randomly fluctuating events, the uncertainty in the number of events is given by the square root of the number of events.
- The uncertainty in the mean value of a set of n measurements that have a standard deviation of s_n is:

$$\sigma_m = \frac{s_n}{\sqrt{n}}$$

- The rules for combining uncertainties in sums, differences, products, ratios and powers are given in Table 13.4.
- For the specific example of converting uncertainties in flux measurements to uncertainties in astronomical magnitudes:

$$\delta m = 2.5 \log_{10}\left(1 + \frac{\delta F}{F}\right)$$

- When determining the overall uncertainty (or *noise*) in an astronomical flux measurement, the contributions from the flux of the target itself, the *sky noise*, the *thermal noise* of the CCD and the *readout noise* of the CCD should each be taken into account.

QUESTION 13.1

Ten measurements were made of the magnitude of a quasar, and the values obtained were:

$$m_v = 22.0, \ 21.6, \ 21.8, \ 22.3, \ 22.1, \ 22.0, \ 21.9, \ 22.2, \ 21.9, \ 22.2$$

(a) What is the mean value of the quasar's magnitude?

(b) Use the spread of the measurements to estimate the random uncertainty in an individual measurement of the quasar's magnitude.

(c) Calculate the standard deviation of the ten measurements, and compare it with the estimate of the random uncertainty obtained in part (b).

(d) Calculate the uncertainty in the mean magnitude.

QUESTION 13.2

Ten measurements are made of the wavelength of a spectral line in the spectrum of a star. The mean value of these measurements is 585 nm and their standard deviation is 6 nm.

(a) What uncertainty should be quoted for the mean wavelength?

(b) Are the measurements consistent with a true value for the wavelength of 591 nm?

(c) If the mean value needed to be known with a precision of ±1 nm, how many measurements of the wavelength would have to be made?

QUESTION 13.3

In this question we return to the measurements reported in Question 6.2, namely an aperture of radius 12.0 pixels is placed around a star on a CCD image and encloses a total count of 2.50×10^6 photons. An annulus of inner radius 18.0 pixels and outer radius 24.0 pixels surrounds the star and includes a total background count of 7.00×10^5 photons. What is the uncertainty in the instrumental magnitude of the star due to counting statistics? (You may neglect the readout noise.)

14 ANALYSING EXPERIMENTAL DATA

After making various observations, processing astronomical images, taking measurements from the images and estimating uncertainties associated with those measurements, the data then usually have to be processed in some way to obtain the final result. You may think that this part of an investigation is rather trivial compared with setting up the equipment and making the measurements, but carelessness at this stage can ruin all your previous hard work. So here are a few hints that should help you to analyse your data with a minimum of effort and a maximum chance of ending up with an appropriate answer.

(a) **Have you recorded all data necessary for the analysis?** This first point is a reiteration of what was said in Chapter 12 about keeping records in your observatory notebook. It is vital that you record all of the necessary data and information that you will need to calculate the final result. If you leave the observatory, or dismantle the apparatus, and then discover that you omitted to record a vital piece of data needed to analyse your results, your whole investigation may be worthless. In this context, it is worth saying that a clearly laid out notebook will make it much easier for you to check what you have, and what you haven't, recorded.

(b) **Plan your data analysis**. Before you start to do any calculations with your data, think carefully about how you will calculate the final result. You may need to combine algebraic equations, and rearrange them so that the quantity that you want to calculate is the subject of the equation. After doing this, it is advisable to check that the expression that you've derived is correct. One way to do this is by checking the consistency of the units on either side of the equation. Alternatively, you can ask yourself whether the dependence of the subject of the equation on the other variables is what you would expect.

It's a good idea to think about how you will analyse the data when you are planning the measurements that you will make. This will allow you to leave appropriate space in your observatory notebook for extra columns in data tables, if they are needed for the analysis stage. It is also worthwhile developing the habit of laying out calculations neatly, and spacing them out. You are much less likely to make errors if your results are easy to read.

(c) **Substitute numerical values at the end of the analysis**. It is good practice to leave the substitution of numerical data into equations until the latest possible stage of a calculation.

(d) **Check your calculations**. The use of calculators and computers reduces the time and tedium required to analyse data, and allows many methods of data analysis to be used routinely. However, it is important to check the results of calculations, as it is incredibly easy to key an incorrect number into a calculator or to make a simple slip in mental arithmetic. One way to check a calculation is to repeat it in a different order. For example, if you calculated the value of the expression

$$\frac{6.63 \times 10^{-34} \times 2.998 \times 10^{8}}{542 \times 10^{-9} \times 1.602 \times 10^{-19}} = 2.29$$

in the order

$$6.63 \times 10^{-34} \times 2.998 \times 10^{8}/542 \times 10^{-9}/1.602 \times 10^{-19}$$

then try it again in the order

$$2.998 \times 10^8/1.602 \times 10^{-19} \times 6.63 \times 10^{-34}/542 \times 10^{-9}$$

Another check that you can do is to round the numbers in the expression and calculate the value of this approximate expression; it shouldn't differ from your previous calculation by more than a factor of about 3. So with the example above:

$$\frac{7 \times 10^{-34} \times 3 \times 10^8}{500 \times 10^{-9} \times 2 \times 10^{-19}} = \left(\frac{21}{1000}\right) \times 10^{(-34+8+9+19)} = 2.1$$

It is always worth asking yourself whether the result that you have calculated seems reasonable. Of course, there will be occasions when you don't know what a reasonable value would be. Few astronomers would remember the value of the magnitude of the tenth brightest star in the Pleiades. However, if a calculation of the orbital period of Io around Jupiter gave a value of 10^5 years, alarm bells should ring and you should check the calculation and the units!

(e) Retain an appropriate number of digits in sequential calculations. When you are performing a series of calculations, then it is important to record and carry forward an appropriate number of digits at each stage. The number of significant figures in your measured data should be such that the uncertainty that you quote affects the last digit (or possibly the last two digits, if you are quoting the uncertainty to two significant figures because it starts with a '1'). A calculator will generally work to more digits than you require, so in sequential calculations you will not lose any precision. However, if you round an intermediate result and record it in your notebook, and then use this rounded value in subsequent calculations, then you may lose some of the hard-won precision of the original data. It is therefore advisable to carry forward to later stages of a calculation *two more* digits than you are justified in quoting in your intermediate results. At the end of the data analysis, when you have worked out the overall uncertainty, you can round the final result to the appropriate number of significant figures.

(f) Graphs. The results of experiments can often be analysed most effectively using graphical methods, and we will discuss these in the next section.

15 MAKING USE OF GRAPHS

Astronomers frequently use graphs to represent the results of their observations and measurements. Although tables of data and graphs may both be used to summarize the results of the same set of measurements, graphs have a great advantage as visual aids. The form of the relationship between measured quantities, the typical uncertainties in measurements and the presence of anomalous measurements are often readily apparent from graphs, and less so from tables of data. In addition, graphs allow straightforward averaging of experimental measurements, interpolation between measurements and (in simple cases) determination of the equation relating measured quantities.

15.1 Graph plotting: a checklist of good practice

There are a number of useful guidelines to good practice that are worth keeping in mind when you are plotting graphs of observational or experimental data. We will summarize them briefly below.

(a) Plot the independent variable (the one you have control of) along the horizontal axis and the dependent variable along the vertical axis. Plotting graphs in this way is purely a matter of convention, but it is a convention that is almost always followed in science.

(b) Label each axis with the name (or symbol) of the plotted quantity divided by its units. We are then plotting pure numbers on the graph (see Box 12.1 on page 96). In some cases it is helpful to also include powers of ten in the axis label to avoid a lot of zeros or powers of ten with each of the numbers on the scale. So if you had to plot values of distance that were typically 2×10^{10} m, you could label the axis 'distance/10^{10} m' and then label the appropriate point on the scale as '2'. This would be preferable to labelling the axis 'distance/m' and labelling the scale point '2×10^{10}'.

The next three points (c to e) really only apply to graphs plotted by hand, as computer plotting packages will generally scale the graphs appropriately for you. Nonetheless it is worth noting the following.

(c) Choose the ranges of the scales so as to make good use of the graph paper. You should try to avoid cramming all the points into one corner, or even one half, of the graph (Figure 15.1, overleaf). In some situations, it may not be necessary for the scale to go to zero on one or both of the axes. However, bear in mind that it is easy to misinterpret a graph if you don't notice the suppression of the zero of an axis – as politicians and advertisers know only too well!

(d) Choose a scale that makes plotting simple. Scales in which ten small divisions of the graph paper are equal to 1, 10, 100, or any other power of ten are easiest to use. However, scales in which ten small divisions equal 2 or 5, or one of these numbers multiplied by a power of ten, are also manageable. Don't make your life difficult by choosing a scale in which ten small divisions represents 3 or 7 as you would take much longer to plot the data, and the chance of plotting points incorrectly would be very much higher!

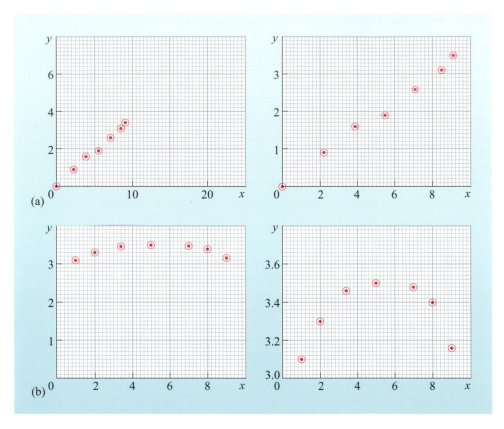

Figure 15.1 (a) These two graphs represent the same data, but the expanded scales in the graph on the right makes better use of the graph paper. (b) Another pair of graphs representing a different set of data from (a). The expanded scale on the vertical axis on the right-hand graph, which starts at 3.0 rather than 0, makes better use of the graph paper.

(e) Plot points clearly. Tiny dots may be confused with dirt on the graph paper (or vice versa) and large dots make it difficult to read off the plotted values from the graph. Either small crosses × or dots with small circles around them ⊙ are recommended. If you are plotting a number of different sets of experimental data on a single graph then it may be clearer if you use a different symbol for the data points for each of the three sets. You may find that it is worth plotting points using a pencil initially, so that they can easily be erased if plotted incorrectly.

(f) Plan your measurements to get a sensible spread of data points on the graph. On a straight line graph this usually means having the data points evenly spaced. However, if you need to determine an intercept, you may want additional points close to the axis. On a non-linear graph, it is usually advisable to concentrate your measurements in regions where the dependent variable changes rapidly. For example, if your graph shows a pronounced peak, you will need to make more measurements in the region where the dependent variable is increasing or decreasing rapidly – the steep slopes on the sides of the peak.

(g) Whenever possible, plot a graph of your results while you are collecting the data. By plotting your data as you collect it, you will immediately notice whether your data points are sensibly spread out. You will also notice if any points are inconsistent with the general trend, and this might be a cue to repeat some of the measurements. It is sensible initially to use a pencil to label the axes and scales and to plot points, since there will be occasions when you decide that a different scale is more appropriate.

Plotting as you go will not always be possible. If you are measuring an effect that changes rapidly with time then there may not be time between the measurements to plot the data. However, even in situations like this, it may be possible to plot-as-you-go if two or more people are collaborating on the experiment.

(h) Draw a straight line or a smooth curve to represent the general trend of the points on the graph. It is seldom appropriate to represent the results of an astronomical measurement by a zigzag line connecting the data points. However, rather than drawing a curve to represent a trend, you may wish to draw a theoretical curve on the graph so that you can compare the data from your experiment with a theoretical prediction. You can do this by calculating pairs of coordinates from the theoretical relationship between the variables, plotting these theoretical points on the graph and drawing a straight line or a smooth curve through them preferably in a different colour from the data. Be careful that the theoretical points are easily distinguished from the experimental data.

For the sake of completeness and for easy reference, we list below a number of other guidelines that we will discuss in more detail later in this book.

(i) Use uncertainty bars to represent the range of uncertainty of the points plotted. If you have been able to estimate the uncertainties associated with your data, then you should represent these by horizontal and vertical bars attached to the plotted points. Uncertainty bars, or error bars as they are often called, are discussed below.

(j) The best-fit straight line or curve should go through most of the uncertainty bars. Also, there should be roughly the same number of points above the best-fit line as below it (see below).

(k) Whenever possible, plot the data in such a way that it can be represented by a straight-line graph. The gradient and intercept of the line can be deduced from a straight-line graph (see below), and these values allow you to write down an equation relating the plotted variables. When drawing graphs by hand, it is much easier to find the best straight line that fits data than to find the best curve, since you can fit a straight line by eye using a transparent ruler. When drawing the best line, think carefully about whether the line should go through the origin. Don't *force* the line through the origin if that is not consistent with your data.

It is also worth noting that although a computer can easily fit an arbitrary function (a curve) to a set of data, this is not always useful. Almost *any* set of data (however poor) can be fit with some sort of mathematical function! However, the function chosen must be consistent with the physics of the situation under study.

15.2 Uncertainty bars

When plotting a graph, you should use **uncertainty bars** whenever possible to indicate the uncertainties associated with your measurements. Both horizontal and vertical uncertainty bars should be plotted, with one set omitted only if the associated uncertainty is too small to show up on the graph.

For example, imagine that you obtained some observations of Cepheid variable stars as a result of which you measured the period of pulsation and the mean magnitude for each. For a particular star, suppose that you determined the period to be $P = 5.6$ days and the mean apparent magnitude to be $m_V = 7.4$. In addition, suppose you estimated that the uncertainty in measuring the period was ± 0.4 days and the uncertainty in the mean magnitude was ± 0.1. The correct way to plot uncertainty bars for this result is shown in Figure 15.2. The circled points indicate the measured values. The uncertainty in each magnitude is indicated by a vertical uncertainty bar, which extends 0.1 above and 0.1 below the circled point, while the period uncertainties are indicated by horizontal bars, which extend 0.4 on either side of the plotted period value. However, suppose that the uncertainty in the period had been ± 0.01 days. Then the horizontal uncertainty bars would have been omitted because they would only be one-tenth of a small division, which is too small to plot on the graph in Figure 15.2.

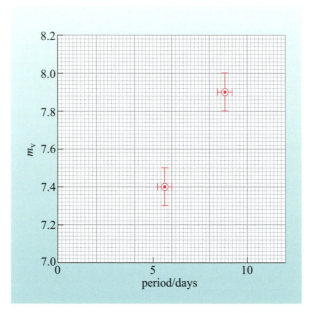

Figure 15.2 Use of vertical and horizontal bars to indicate uncertainties in data. The horizontal bars indicate that there is an uncertainty of ± 0.4 days in the period measurement, and the vertical bars indicate an uncertainty of ± 0.1 magnitudes in the mean brightness measurements.

Plotting uncertainty bars is slightly more complicated if you are not plotting a measured quantity directly, but some function of it, such as A^2, $\sin A$ or $\log A$. In such cases, you will need to use some of the rules that were quoted in Table 13.4. For example, if a power is involved, you can use:

percentage uncertainty in $A^n = n \times$ percentage uncertainty in A

Thus if the uncertainty in A is $\pm 5\%$, the uncertainty in A^2 is $\pm 10\%$.

For other functions $f(A)$, the most straightforward procedure is to plot the maximum estimated value $f(A + \delta A)$ and the minimum estimated value $f(A - \delta A)$ and draw the uncertainty bar to join them. For example, if $A = (2.0 \pm 0.1)$, and you want to plot log A with its error bar, then you would plot a point at $\log_{10} 2.0$ (= 0.30), and draw an uncertainty bar to span the range from $\log_{10} 2.1$ (= 0.32) to $\log_{10} 1.9$ (= 0.28). Note that the uncertainty bars will vary from point to point if $f(A)$ is not a linear function (a straight line), even if the uncertainties in the original measurements of A are all the same, and that the uncertainty bars may be asymmetric.

Plotting uncertainty bars on a graph serves a number of useful purposes. The relationship between the plotted quantities can often be represented by a line that is either straight or a smooth curve, and we would then expect the data points to be scattered fairly equally on either side of that line. If the uncertainty bars represent the standard deviation of the plotted quantities, then we would expect the line to pass through about 2/3 of the bars. If the measured values deviate from the line by much more than the uncertainty bars, as shown in Figure 15.3a, then either we have underestimated the uncertainties, or the assumption that this line describes the results is not valid. On the other hand, if the line passes close to the centre of all of the uncertainty bars, as in Figure 15.3b, then it is likely that we have overestimated the uncertainties. An alternative explanation in this latter case could be that the dominant contribution to the uncertainty bars is from a systematic uncertainty, so that the 'true' curve could be similar to the curve drawn in Figure 15.3b but shifted upwards or downwards on the graph within the range of the uncertainty bars.

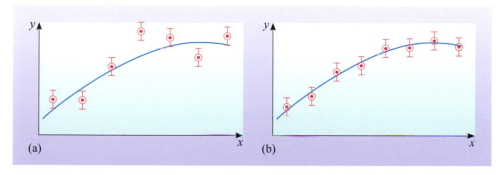

Figure 15.3 (a) Have the uncertainties in the data been underestimated? (b) Have the uncertainties been overestimated?

Uncertainty bars are also very helpful in identifying mistakes in measurements or in plotting the points. For example, if a point deviates from the general trend of the other data points by significantly more than the size of the uncertainty bars, it would be reasonable to be suspicious of the 'odd' measurement. In cases like this, the plotting of the anomalous point on the graph should be checked, and any calculations made to derive the plotted values should be checked as well. If these checks don't show up a mistake, then the measurements that gave rise to the suspect point should be repeated. In some cases, of course, repeating the measurements is just not possible, and one is left with the difficult decision about whether to ignore the point or not. No hard and fast rule can be made about this, but you should not discard data lightly. Bear in mind that what appears to be an anomalous measurement may actually indicate a real (and possibly undiscovered) effect!

15.3 Straight line graphs: gradients, intercepts and their uncertainties

Astronomers frequently plot their data in such a way that they can be represented by a straight line. One reason for this is that deviations from a straight line are much easier to see than deviations from a curve. However, just as important, it is easy to deduce the equation that represents a straight line that you have drawn on a graph. This equation expresses the relationship between the experimental quantities, and it may be possible to compare the values of the numerical constants in the equation with those predicted by a theoretical relationship.

The equation of any straight line on a graph of y versus x can be written as

$$y = mx + c \tag{15.1}$$

The significance of the constants m and c in this equation is illustrated in Figure 15.4a. The constant m is the **gradient** of the line:

$$\text{gradient } m = \frac{\text{rise}}{\text{run}} = \frac{\Delta y}{\Delta x} \tag{15.2}$$

and c is the **intercept** with the y-axis, i.e. the value of y when $x = 0$. Thus

$$\begin{pmatrix} \text{value plotted on} \\ \text{vertical axis} \end{pmatrix} = \text{gradient} \times \begin{pmatrix} \text{value plotted on} \\ \text{horizonal axis} \end{pmatrix} + \begin{pmatrix} \text{intercept with} \\ \text{vertical axis} \end{pmatrix} \tag{15.3}$$

Figure 15.4b shows a sketch graph of (instrumental magnitude – catalogue magnitude) versus airmass for a set of standard stars measured during a night's observing. By analogy with Equations 15.2 and 15.3, the equation for the straight line plotted there is:

$$m' - m = (\text{gradient} \times X) + (\text{intercept with vertical axis}).$$

The gradient of this line is given by

$$\text{gradient} = \frac{\text{rise}}{\text{run}} = \frac{\Delta y}{\Delta x} = \frac{\Delta(m' - m)}{\Delta X}$$

Figure 15.4 (a) A graph that illustrates the meaning of the constants m and c in the general equation of a straight line, $y = mx + c$. (b) A straight-line graph that represents (instrumental magnitude – catalogue magnitude) versus airmass for a set of standard stars measured during a night's observing.

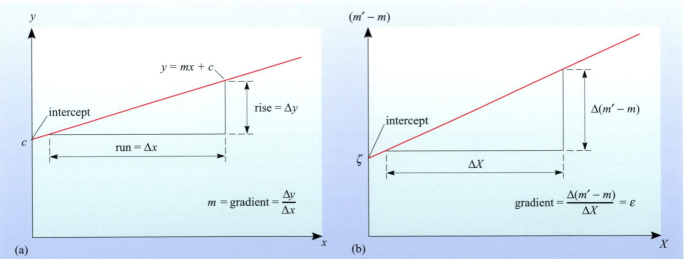

(a)

(b)

and this is equal to the extinction coefficient (in magnitudes per airmass) for a particular filter. The intercept with the vertical axis represents the *zero-point offset* of the magnitude calibration.

Now consider a set of observations in which the (instrumental magnitude – catalogue magnitude) for a range of standard stars is measured at various airmasses. A graph of the observational data is shown in Figure 15.5.

■ What can you deduce, from the graph in Figure 15.5, about the uncertainties in the measurements of magnitudes and airmasses?

❏ The vertical uncertainty bars are all the same size, and correspond to an uncertainty in the magnitude measurements of ±0.05. No horizontal uncertainty bars are shown, which implies that the uncertainty in the airmass was less than one division on the graph paper, i.e. less than ±0.01.

■ Are the data consistent with a constant value of the extinction coefficient throughout the night?

❏ Figure 15.5 shows that a straight line can be drawn that passes through all of the uncertainty bars, and this means that the data are consistent with a linear relationship between (instrumental magnitude – catalogue magnitude) and airmass. (In fact the uncertainties may have been overestimated a little.) The data are therefore consistent with a constant value of the extinction coefficient, since a linear relationship between (instrumental magnitude – catalogue magnitude) and airmass is precisely what is expected for a constant extinction coefficient (see Chapter 6 Section 4). Of course, if we had been unable to draw a line that passed through most of the uncertainty bars, then we would have concluded either that the extinction coefficient was not constant, or that the uncertainty in some of the measurements was greater than indicated.

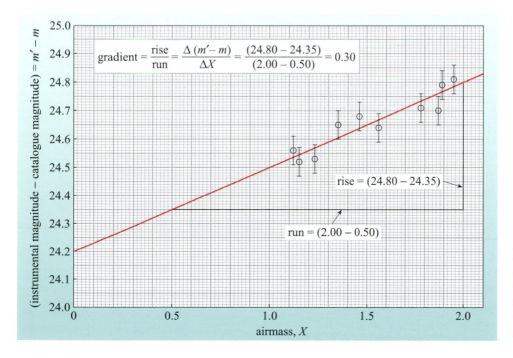

Figure 15.5 A graph showing measured values of the (instrumental magnitude – catalogue magnitude) for a range of standard stars, measured at various airmasses.

Now let's make some quantitative deductions from the straight-line graph in Figure 15.5. The gradient of the straight line in Figure 15.5 is our best estimate of the extinction coefficient. To work out the gradient of the line, we select any pair of points *that lie on the straight line*. (Note that the straight line effectively averages out the uncertainties in individual data points, so we always work with values on the line, rather than with values from the original data, which may well not be on the line.) The rise $\Delta(m' - m)$ is then the difference between the $(m' - m)$ values of the two selected points, and the run ΔX is the difference between the X-values of these points, as shown in Figure 15.5. Thus

$$\text{gradient} = \frac{(24.80 - 24.35)\,\text{magnitudes}}{(2.00 - 0.50)\,\text{airmasses}}$$

$$= \frac{0.45\,\text{magnitudes}}{1.50\,\text{airmasses}}$$

$$= 0.30\,\text{magnitudes per airmass}$$

Note that the gradient does *not* depend on which pair of points on the line you use to calculate it. However, the coordinates of the points can be estimated from the graph with only limited precision. The best practice, therefore, is to choose two points on the line that are as widely separated as possible. The uncertainties in reading the coordinates will then be much smaller fractions of the differences between the coordinates than if the two points were close together.

The straight line drawn in Figure 15.5 is our choice of the 'best fit' to the data. It was drawn so that it passes as close as possible to the data points and the points above the line are balanced by other points below the line. Remember that we expect the line to pass through about 2/3 of the uncertainty bars. However, many other lines with different gradients could also be drawn to pass through about 2/3 of the uncertainty bars. In order to obtain an estimate of the *uncertainty in the gradient*, we can draw the line that has the *maximum* possible gradient consistent with the data, and similarly draw the line that has the *minimum* possible gradient consistent with the data. These lines are shown in Figure 15.6, and their gradients are 0.41 magnitudes per airmass and 0.19 magnitudes per airmass, respectively. These values differ from the gradient of the 'best' straight line by ±0.11 magnitudes per airmass. We can quote the value of the gradient as (0.30 ± 0.11) magnitudes per airmass.

The other quantity that specifies the position of a straight line on a graph is its intercept with the vertical axis, and we have called this ζ, the *zero-point offset* of the magnitude calibration. For the straight line shown in Figure 15.5, the intercept is at $\zeta = 24.20$ magnitudes and, since this is determined from the 'best' straight line, it is the best estimate of the intercept. The upper and lower limits for the intercepts are found, as you might expect, from the intercepts of the lines with minimum and maximum gradients shown in Figure 15.6. These intercepts are 24.06 and 24.39 respectively, and they differ from the best intercept by +0.19 magnitudes and –0.14 magnitudes, respectively. The average of these two numbers, rounded to two significant figures, is 0.17 magnitudes. The experimentally determined intercept can therefore be quoted as (24.20 ± 0.17).

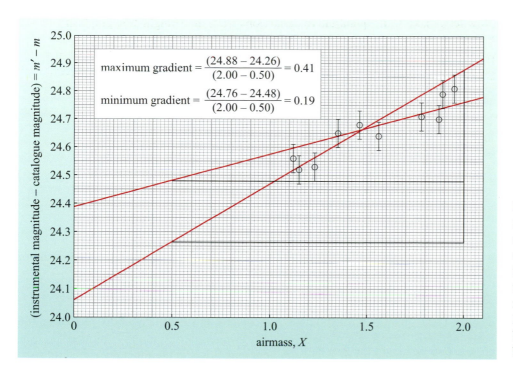

$$\text{maximum gradient} = \frac{(24.88 - 24.26)}{(2.00 - 0.50)} = 0.41$$

$$\text{minimum gradient} = \frac{(24.76 - 24.48)}{(2.00 - 0.50)} = 0.19$$

Figure 15.6 The uncertainty in the gradient is estimated by drawing the steepest and shallowest lines that still pass through about 2/3 of the uncertainty bars, and then measuring the gradients of these lines. (These lines would normally be drawn on the same graph as the best fit line.)

We can use the values for the gradients and intercepts that we have determined from the graphs in Figure 15.5 and Figure 15.6 to write down an equation for the relationship between (instrumental magnitude − catalogue magnitude) and airmass for this particular calibration:

$$m' - m = (0.30 \pm 0.11) \text{ magnitudes per airmass} \times X + (24.20 \pm 0.17) \text{ magnitudes.}$$

Of course, if you are using a computer graph plotting package, then the software will usually calculate the values and uncertainties in the gradient and intercept of the graph for you. However, it is useful to appreciate what is going on 'behind the scenes' to arrive at such a result.

15.4 Transforming curves into straight lines

You have seen that it is straightforward to deduce an equation relating two variables when their relationship can be represented by a straight-line graph. Can the same thing be done with curved graphs? Unfortunately, the answer is no; it cannot be done directly. However, it is often possible to plot *functions* of the original variables (e.g. s^2, $\sin \theta$, $\log_e N$, etc.) that are represented by a straight line. It is then possible to deduce the equation for this linear relationship from a graph, and hence deduce the relationship between the original variables. We will now give a couple of examples of how this could be done in practice.

First, suppose that we measured the area A and radius r of a set of circles. Plotting A versus r will certainly not produce a straight line: it produces a parabola, as shown in Figure 15.7a. However, if A is plotted against r^2, then the resulting graph *is* a straight line, as shown in Figure 15.7b, with a gradient that appears to be about 3.1 and intercept zero.

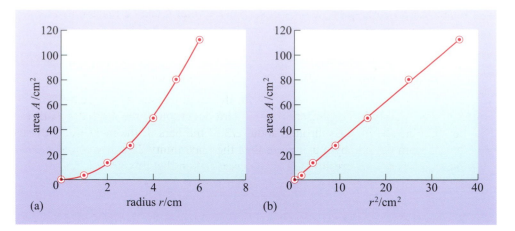

Figure 15.7 (a) Measurements of the area A of circles of radius r. The relationship is parabolic. (b) The same data as in (a), but plotted as area A versus r^2. This graph demonstrates that A is linearly related to r^2.

It is instructive to relate the graph in Figure 15.7b to the general form of the equation of a straight line

$$y = mx + c, \tag{15.1}$$

where y is the quantity plotted on the vertical axis and x is the quantity plotted on the horizontal axis. The area A plotted vertically is analogous to y, and r^2 plotted horizontally is analogous to x. The gradient m is 3.1, and the intercept c is zero, so there is no constant term. The equation of this line is therefore $A = 3.1r^2$, which is consistent with a well-known result from geometry, $A = \pi r^2$.

An alternative way to obtain a straight line is to take logarithms of both sides of the known formula to obtain $\log(A) = \log(\pi) + 2\log(r)$. Clearly, if we plot $\log(A)$ versus $\log(r)$ for the set of circles measured we will obtain a straight line whose gradient is equal to 2 and whose intercept on the vertical axis is at $\log(\pi)$.

Taking logarithms of an equation such as this also enables an unknown equation to be determined. For instance suppose we don't know precisely how the volume V of a sphere depends on its radius r, but we guess the equation is something like $V = pr^q$ where p and q are both unknown constants. Taking logarithms we obtain

$$\log(V) = \log(p) + q\log(r)$$

So a graph of $\log(V)$ versus $\log(r)$ will yield a straight line whose gradient is equal to q and whose intercept is at $\log(p)$. By measuring these from the graph we would soon determine that $p = 4.19$ and $q = 3$, which accords with the well-known formula $V = (4\pi/3)r^3$.

We recommend that whenever possible, you should plot your data in a way that will give you a straight line. Deviations from linearity are easy to see, so a straight-line graph soon shows up any deficiencies in the data.

15.5 Summary of Chapter 15 and Questions

• Rules for good graph plotting include: plot the independent variable (the one you have control of) along the horizontal axis and the dependent variable along the vertical axis. Label each axis with the name (or symbol) of the plotted quantity divided by its units and plot the points clearly. Choose the ranges of the scales so as to make good use of the graph paper and a scale that makes plotting simple. Plan your measurements to get a sensible spread of data points on the graph and whenever possible, plot a graph of your results while you are collecting the data.

- Whenever possible, plot the data in such a way that it can be represented by a straight-line graph and then draw a straight line or a smooth curve to represent the general trend of the points. Use uncertainty bars to represent the range of uncertainty of the points plotted. The best-fit straight line or curve should go through most of the uncertainty bars.

- Data points should be scattered fairly equally on either side of the best-fit line. If the uncertainty bars represent the standard deviation of the plotted quantities, then the line should pass through about 2/3 of the bars. If the measured values deviate from the line by much more than the uncertainty bars, then either the uncertainties are underestimated, or the assumption that this line describes the results is not valid. If the line passes close to the centre of all of the uncertainty bars, then it is likely that the uncertainties are overestimated or that the dominant contribution to the uncertainty bars is from a systematic uncertainty.

- The equation of any straight line of a graph of y versus x can be written as $y = mx + c$, where m is the gradient of the graph and c is the intercept with the y-axis. The best-fit straight line to a set of data allows the values of m and c to be determined directly from the graph.

- Limiting values for the gradient and intercept may be determined by drawing lines corresponding to the maximum and minimum gradients that are consistent with the data, within the uncertainty bars.

- By guessing the functional form of a relationship between two variables, functions of these variables may be plotted on the vertical and/or horizontal axes of a graph in order to render the data in a form that may be fit by a straight line relationship.

QUESTION 15.1

Figure 15.8 shows a graph to determine the extinction coefficient and zero point from a set of photometric standard star observations. What improvements do you suggest could be made to this graph?

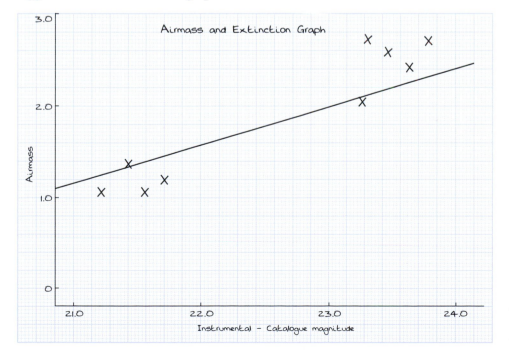

Figure 15.8   A graph containing several examples of bad practice, for use with Question 15.1.

QUESTION 15.2

The orbital periods (P) and mean distances from the planet (a) are measured for several of Jupiter's satellites. The periods and distances should be related by Kepler's third law, $a^3/P^2 = GM/4\pi^2$. Suggest *two* graphs that you could plot, each of which should yield a straight line that confirms the validity of the relationship.

16 USING CALCULATORS AND COMPUTERS

Programmable graphical calculators and computers can often greatly simplify the analysis and display of experimental data. They can be used to calculate the mean and standard deviation of a set of measurements, to plot a graph of data, to determine the best-fit straight line or curve for your data, and much more besides.

It is not appropriate to give specific guidance here on how you can carry out these various operations, because the detailed procedures depend on the particular combination of hardware and software that are available to you. However, we will indicate some of the possible applications, and you can investigate for yourself how to implement them with your calculator or computer.

Spreadsheet programs are very powerful tools for the astronomer, and Figure 16.1 (overleaf) illustrates a few of the ways in which they can be used to display and analyse data. In the example shown there, the data are the measured instrumental magnitudes of a set of standard stars used to determine the extinction coefficient and the zero-point offset for the calibration. Four Landolt standard stars lie in each image frame: the stars are labelled with their catalogue numbers as 107–602, 107–601, 107–600 and 107–599, and the frames are numbered between 85 and 245. A set of 13 frames was obtained at a range of airmasses throughout the night. The catalogue magnitudes of the standard stars are listed at the top of the spreadsheet.

(a) Produce a table of data. You can enter your measurements into the rows and columns of a ready-made table, as shown in area (a) of Figure 16.1. Each number is keyed into a separate box, called a *cell*. Columns and/or rows can be labelled to indicate the quantities tabulated, and their units. In the example in Figure 16.1, thirteen sets of values of airmass and instrumental magnitude have been entered.

(b) Calculate functions of values in one or more cell. In area (b) of Figure 16.1 we have calculated the instrumental magnitude minus the catalogue magnitude for each observation of each standard star. The computer was instructed first to evaluate the value corresponding to the first observation of the first standard star (number 107–602 in frame number 85), and this instruction was then copied to all of the other 'instrumental minus catalogue magnitude' cells in order to complete the data in these columns. There are big savings of time when large data sets are used.

(c) Calculate the mean and standard deviation of a set of data. You can instruct the computer to calculate the mean of the values in a set of cells and enter this in another cell. Similarly, you can instruct the computer to calculate the standard deviation of the values in a set of cells. In Figure 16.1, area (c), the computer has calculated the means and standard deviations for the (instrumental magnitude – catalogue magnitude) of the four standard stars in each frame. It only requires about half a dozen mouse clicks to calculate each of these values, irrespective of how large the data set. This means that once you have familiarized yourself with the procedure, you can save yourself a lot of time when analysing data this way.

Most scientific calculators have a facility for calculating the mean and standard deviation of a series of numbers: you simply key in the numbers, press the 'mean' key and note the result, and then press the 'standard deviation' key and note the result. Again, this is much easier than doing the standard deviation calculation 'by hand' in the way shown earlier.

(d) Plot a graph. The computer can plot a graph of data from two columns (or two rows), and in area (d) in Figure 16.1 a graph of mean (instrumental minus catalogue magnitude) versus airmass has been plotted. You can choose the style of the graph e.g. displaying data points only, or displaying a histogram. Normally the computer will automatically choose appropriate scales to display the data, but you can make your own choices of scale if you wish. Uncertainty bars can be displayed, though in some spreadsheet programs the uncertainty bars have to be either the same size for each point or the same percentage of each value plotted. Here we have calculated the uncertainty in the mean for each set of four standard stars (according to $\sigma_{\mathrm{m}} = s_n / \sqrt{n}$) and plotted this value as the uncertainty bar in each case.

(e) Fit a straight line to the plotted data. In the graph in Figure 16.1, the computer has fitted a straight line to the data, and has displayed the equation of the fitted line. The intercept (24.51) is the *zero-point offset* of the calibration and the gradient (0.18) is the extinction coefficient in magnitudes per airmass, for the V-filter. Note a limitation of this spreadsheet program: it does not display the units of the constants in the fitted equation. When you transcribe equations displayed by the computer, you will therefore have to work out the appropriate units for the constants, and insert them into the equation. Nor will the spreadsheet program generally allow you to fit highest and lowest gradients to the gradient to ascertain the uncertainty limits of the gradient and intercept.

As well as fitting straight lines to data, it is also possible to fit curves described by various other functions, and these functions include polynomials and logarithmic

Figure 16.1 An example of how a spreadsheet can be used for data analysis and display. The areas of the spreadsheet labelled a, b, c, etc. are described in the text.

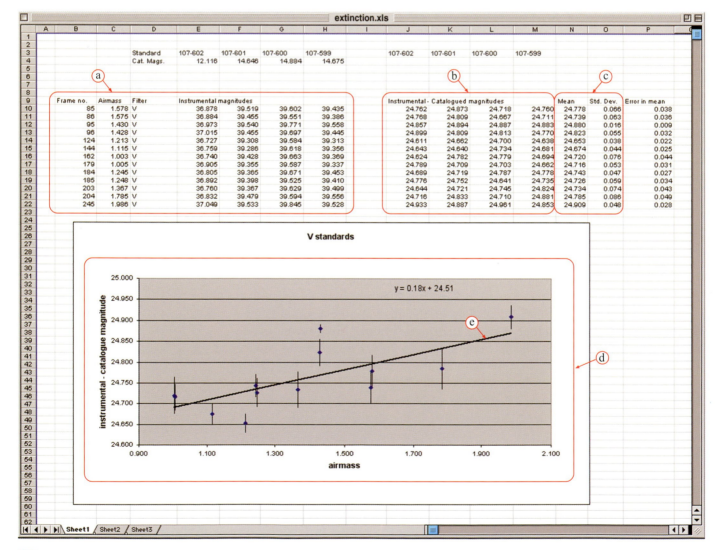

functions. Fitting a straight line or a curve to data normally uses a statistical procedure known as the *least-squares method*, and this is discussed in Box 16.1.

Using the least-squares method on a computer to find the best-fit straight line for a data set is far easier than plotting a graph by hand and then calculating gradients and intercepts. However, it may not be as easy to 'plot as you go' using a computer. You may therefore need to plot a rough graph while taking measurements, and then use a computer for a least-squares fit when all of the data have been collected.

BOX 16.1   THE LEAST-SQUARES METHOD OF CURVE FITTING

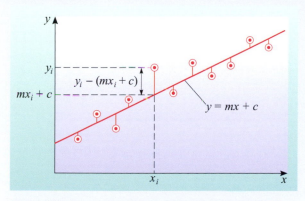

Figure 16.2   A set of data points and a straight line drawn to represent the data. The vertical lines between the data points and the line represent the deviations of the data points from the line.

We will illustrate the least-squares method by considering the fitting of a straight line to the data plotted in Figure 16.2. A variety of straight lines, represented by $y = mx + c$, could be drawn on the graph, and we want to find the values of m and c that produce the line that is the 'best fit' to the data. For simplicity, we will assume that the uncertainties in y are the same for all data points, and the uncertainties in x are negligible.

What do we mean by 'best fit'? Clearly, we want the line to be positioned so that it passes as close to the data points as possible. So we want to find the values of m and c that in some sense minimize the lengths of the short lines shown connecting the data points with the straight line in Figure 16.2. Now if the coordinates of the ith data point are (x_i, y_i), then the y-coordinate of the straight line when $x = x_i$ must be $(mx_i + c)$. Thus the vertical separation between the ith data point and the straight line is just $y_i - (mx_i + c)$, and it is these separations that are represented by the short lines in Figure 16.2.

Now we can't find the best-fit straight line by simply minimizing the sum of all of the separations $y_i - (mx_i + c)$, since the separations are both positive and negative.

Instead we find the best line by minimizing the sum of the *squares* of the separations (which must all be positive), i.e. we minimize

$$\Sigma(y_i - mx_i - c)^2 = (y_1 - mx_1 - c)^2 + (y_2 - mx_2 - c)^2 + (y_3 - mx_3 - c)^2 + \dots$$

This is why the method is known as the **least-squares method**.

Standard techniques of calculus can be used to evaluate the conditions for this summation to have its minimum value. They lead to explicit expressions for the gradient m and the intercept c that appear in the equation of a straight line. We will not reproduce the expressions for m and c here, since the computer or calculator that you use to implement the least-squares method will display the results for m and c automatically.

The least-squares method is not limited to fitting a straight line to data, but can be used to fit a variety of other functions, such as:

- a power law, of the form ax^b,
- a polynomial function; say, $ax^3 + bx^2 + cx + d$; programs generally allow a choice of the highest power included in the function;
- an exponential function; for example, ae^{bx};
- a logarithmic function of the form $a \log_e x + b$.

For each of these functions, the computer program can calculate the values of the constants (a, b, c, d) that minimize the sum of the squares of the separations of the data points from the curve. In many situations, you will base your choice of the type of function to fit to the data on a theoretical relationship between the variables. In other cases, you may be trying to identify the (unknown) functional relationship between the measured variables. You then need to compare the best curves computed for different types of function with the data points to see which curve is the best representation of the data.

17 COMMUNICATING YOUR RESULTS

Scientists need to be able to communicate their ideas and results effectively. There are many examples in the history of science where scientists have made significant discoveries, but not publicized them. For example, Henry Cavendish was an eighteenth century British scientist, who carried out extensive experiments on electrical phenomena. However, he rarely took the trouble to write up his findings, so that for many years other people continued to work on problems that he had already solved. He discovered that the current through a resistor is proportional to the voltage applied across it. Fifty years later Ohm made the same discovery. Yet, this relationship is now known as Ohm's law, because it was Ohm who made his discovery known to others.

If communication of scientific results was important in the eighteenth century, it is perhaps even more essential in the twenty-first century. A few centuries ago, science was only carried out by a small number of people, but now it is a huge industry, spread over academic and industrial institutions all round the world. The progress of science is very much a cooperative effort, with one group of scientists building on the discoveries of others, and then communicating their own results for other scientists to build on in turn. So research scientists publish papers describing their work in academic journals, and they attend conferences at which they give formal talks and have informal discussions about their latest findings.

Skill in the presentation of experimental results and the conclusions derived from them is something that every astronomer should acquire, since the acceptance or rejection of their results and conclusions will usually be based on their written or oral reports. As a student, you will be able to develop and practise this skill by writing reports of experiments and possibly by giving oral presentations of your results to other students. If you fail to communicate clearly your results and their significance, your work is unlikely to get the recognition that it may deserve. What's more, if you do not disseminate your work, then it might as well not have been conducted because no one else will benefit from it.

In astronomy, there are well-established organizations for linking amateurs' results to professional research, such as the British Astronomical Association (BAA) and the American Association of Variable Star Observers (AAVSO). So it is by no means unusual for amateur astronomers to have their results published.

17.1 Writing reports

There is a big difference between the way that you record information in your observatory notebook, which is your personal record of your work, and the way in which you present your final report. The main requirement of your notebook is that it should contain a record of all necessary information about the work you have carried out in a form that will be accessible and intelligible to you in six months time, as well as immediately after completing the work. It should also allow another student or a colleague to repeat what you have done. A final report must contain all of the necessary information, but it must be presented in a way that is accessible to, and readily understandable by, the intended readers. Clarity, conciseness, and coherence of presentation are all at a premium. Any conclusions that you reach should be stated simply, with a clear indication of their limitations.

Here are a few guidelines for the presentation of reports. They will give you an idea of what is likely to constitute an acceptable framework for a report, but you should not regard them as compulsory in all cases.

(a) Give your report a title. The title should be brief, but it needs to give the reader a clear idea of the subject of the report. If the aim of the investigation was to measure a particular quantity, to observe a particular phenomenon, or to confirm a theoretical prediction, then state this explicitly in the title. Novelists often choose titles that sound intriguing but give little idea of what their books are about; scientists should aim to provide a title that conveys relevant information.

(b) Start with an abstract. The first component of your report should be an abstract that summarizes the work in a few sentences. It is usually best to write the abstract *after* you have completed the rest of the report, but it should appear at the beginning.

The important function of the abstract is to succinctly provide information that will allow a reader to decide whether they are interested in reading the report. It therefore needs to give the reader an overall picture of the general scope of the report and of the final results and main conclusions. Since it will be the first part – or perhaps the only part – of the report that a reader will see, the abstract needs to be intelligible without reading the main body of the report. In particular, you shouldn't use undefined terms or symbols with which the reader may not be familiar.

To illustrate the importance of abstracts, consider the astronomy E-print archive hosted by arXiv.org. This may be found at the web address

http://uk.arxiv.org/archive/astro-ph

On a typical day, around 25 astronomy articles are submitted to this archive – that's around 750 a month or of order 10 000 every year. No practising astronomer can hope to read 25 complete journal articles every day (including weekends!) However, what most people can do is scan through the titles each day of those papers submitted in the previous 24 hours, and then read the abstracts of those that appear to be relevant to one's own research. If the abstract itself is convincing, then and only then might the astronomer download the complete article from the archive and read that.

(c) Introduction. In the introduction you should briefly set out the purpose of the work you have carried out, and give a brief outline of how it was done. It may be appropriate to describe the theoretical background in order to put the work in context, but this may not always be necessary. If the purpose of the work is to verify a relationship or measure a specific quantity then you will need to explain what these are. Sometimes it may be appropriate to make references to textbooks or other documents where relationships are derived.

(d) Method. This section is the place where you describe how you carried out the investigation. You need to include details of the equipment used, any special precautions that you took to reduce uncertainties, any checks that you made and any problems encountered.

If you are writing a report of a project in which you have followed a procedure described in student notes, then there is no need to repeat all of the details that are in the notes. A brief summary and a reference to the notes will generally be enough.

(e) Results. The results section is the place to present the measurements or observations that you have made. It is worth taking some trouble to organize your data into an easily digestible form, and this is often done most neatly using tables and graphs. You should make sure that each of these has a caption, and that it is referred to at the appropriate place(s) in the text. It is essential to label clearly the columns (and/or rows) in tables and the axes of graphs, and to include the correct units.

This section should also contain the analysis of the data and the calculation of the result. It is not necessary to show all of the steps in the derivation of algebraic expressions that you use to analyse your data, unless you are using a non-standard method. Estimates of uncertainties in measurements should be included, together with an indication of how the individual uncertainties in measured values were combined to obtain the uncertainty in the final result.

(f) Discussion. This is where you discuss how your work and the result fit in with other work and relevant theories; for example, is the result consistent with other published values? Make sure that you discuss any unexpected behaviour or results. You should also point out any assumptions or approximations that you have made, limitations in the procedure or the analysis, and you might suggest ways in which the investigation could be improved. You may also have suggestions for additional projects or further investigations arising from the work that you have done.

(g) Conclusion. This is usually the final part of the account, and it encapsulates in a few sentences the main outcomes of the work. It may include a numerical result – which should always be accompanied by an estimate of the uncertainty – or an equation that you have deduced or verified.

17.2 Good scientific writing

This is not the place to give detailed advice on how to write good English, but here are a few pointers that may be useful when you are producing reports of your work.

(a) Think about the structure of your report before you write it. Dividing your report into the sections that we have just described is a good way to provide the main structure that is required. You then need to decide on the best order for the components of each section, and you may find it helpful to jot down an outline of the main topics that you need to include. This will help you to think about the links between the various topics and to fix on an order that seems logical. If there is a logical flow to the report, it will be much easier for you to write, but, more important, it will be much easier for the reader to follow!

(b) Avoid very long sentences. It is possible to write clearly using long sentences, but it is much more difficult than writing short sentences. However, stringing together a series of very short sentences is likely to result in a rather boring prose style. Varying the length of your sentences will make your report more interesting to read.

(c) Divide long sections into paragraphs. Start a new paragraph when you start a new topic or a different aspect of a topic. This makes your writing easier for the reader to follow. The first and last sentences in a paragraph often have the most impact on the reader, and some writers try to put their main points in these places.

(d) Active or passive voice? Opinion is divided over whether reports and scientific papers should be written in the active voice ('I measured the magnitude of the star

…', or 'We measured the magnitude of the star …') or the passive voice ('The magnitude of the star was measured …'). Your aim should be to make your written reports objective, but this doesn't mean you have to write in the passive voice. Most authors avoid the use of 'I' in scientific papers, but this is certainly not a rigid rule. Using 'we' is certainly acceptable, particularly when describing the actions or views of more than one person. Use of the passive voice to convey a sense of (unjustified) authority should be avoided. In particular, you should clearly distinguish between your own interpretations and those that are more widely accepted: thus writing about 'our interpretation of these results …' or 'we suggest that these results show …' is often preferable to 'the results show that…'.

(e) Read your own writing critically. If you find that you are having difficulty expressing something clearly, a possible approach to use is to ask yourself what you would say if you were telling somebody about this subject. Some people find it useful to read aloud to themselves what they have written. This helps to give them a feel for the flow of the words; usually if something is difficult for you to read to yourself when you know the intended meaning, then it will be difficult for your readers to follow. You can read to yourself as you complete a sentence, a paragraph, or a section. It is also very helpful to put aside what you have written for a couple of days, and then come back to it. Reading your written work through afresh will often make you aware of gaps in the logical flow of topics, or aware of points that are missing or unclear. It is also very useful if you can persuade someone else to read it and give you critical comments.

(f) Using a word processor can often be very helpful. The advantage of a word processor is that it is very easy to change the wording, change the order for topics, change the paragraphing, and so on. Some people like to dash off a quick draft, and then polish this up, while others like to aim for something that is fairly polished first time round. Whatever your preference, a word processor gives you the freedom to modify and polish your draft bit by bit without having laboriously to rewrite the whole thing from scratch. It also enables you to check and correct spelling, and possibly your grammar too.

ANSWERS AND COMMENTS

QUESTION 1.1

The stars are very nearly in opposite directions in the sky: the declinations have nearly equal magnitude and opposite signs, and the right ascensions are nearly 12 hours different. Hence they are roughly 180° apart.

QUESTION 1.2

(a) (i) The altitude of the north celestial pole, for an observer in the Northern Hemisphere, is the observer's latitude *lat* (see Figure 1.8). (ii) The zenith is vertically overhead, and so the altitude is +90° (see Figure 1.9).

Comment: Negative altitudes are below the horizon.

(b) The angle of the Sun from the zenith is +90° 0′ − 64° 21′ = 25° 39′.

QUESTION 1.3

At latitude 90° N, the celestial equator is the horizon; see Figure 1.21a.

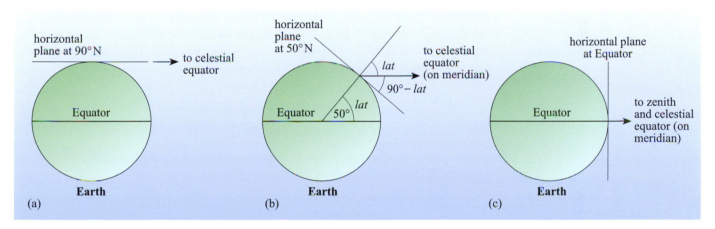

(a)　　　　　　　　　　(b)　　　　　　　　　　(c)

At latitude 50° N, the celestial equator meets the horizon at the east and west points (Figure 1.7). The celestial equator lies in the south, reaching a maximum altitude at the meridian, where the altitude is 90° − 50° = 40° (Figure 1.21b).

Figure 1.21 See the answer to Question 1.3.

At latitude 0°, i.e. the Equator, the celestial equator runs through the east and west points, and the zenith (Figure 1.21c).

Note: At *any* latitude away from the poles, the celestial equator meets the horizon at the east and west points (see Figure 1.7) and reaches a maximum altitude at the meridian, given by (90° − *lat*).

QUESTION 1.4

(a) The scale around the edge of the planisphere shows that 27 December is aligned with 23:00 hours. Hence the picture shows the night sky as visible on 27 December at 11 p.m.

(b) The constellations near the western horizon are Pegasus and Pisces, hence stars in these two constellations are in the process of setting.

(c) Extending the meridian line to the edge of the planisphere in the south, shows that it corresponds to *RA* ≈ 5.5 hours at this time.

(d) The most southerly declination visible at this time is also to the south. Circles of constant declination are shown in the window at +60°, +30°, 0° (the celestial equator) and −30°. Part of this latter circle is visible in the south, from which it may be estimated that the most southerly declination visible is about −35°. (In fact, since the planisphere is designed to operate at latitude 51.5° N, the most southerly declination visible is 90° − 51.5° = 38.5°.

QUESTION 2.1

(i) The *light-gathering power* is proportional to D_o^2, but independent of f_o. (ii) The field-of-view is inversely proportional to f_o, and independent of D_o, although it does depend on the diameter of the eyepiece field stop. (iii) The angular magnification is proportional to f_o, but independent of D_o. (iv) The limit of angular resolution is inversely proportional to D_o and independent of f_o, but note that it varies with wavelength.

QUESTION 2.2

(a) Light-gathering power is proportional to (aperture diameter)^2, so the ratio of light-gathering powers for the two telescopes is $(5.0/1.0)^2 = 25$.

(b) The (theoretical) limit of angular resolution is inversely proportional to the aperture of the objective lens or objective mirror. Thus, a telescope with $D_o = 5$ m can theoretically resolve two stars with an angular separation five times smaller than a telescope with $D_o = 1$ m (neglecting air turbulence and aberrations). In practice, of course, for ground-based telescopes, atmospheric seeing is usually the limiting factor.

QUESTION 2.3

(a) The aperture of a *diffraction-limited* telescope that would have a resolving power of 1″ at a wavelength of 485 nm is given by

$$D_o = 1.22\lambda/\alpha_c$$

Now 1″ is equivalent to $\pi/(180 \times 3600)$ radians $= 4.85 \times 10^{-6}$ radians. So

$$D_o = (1.22 \times 485 \times 10^{-9} \text{ m})/(4.85 \times 10^{-6})$$

$$= 0.122 \text{ m}$$

A *diffraction-limited* telescope with an aperture diameter of only 12 cm would therefore have an angular resolution of 1″ when operating at a wavelength of 485 nm (i.e. in the blue part of the visible spectrum).

(b) Although a large (and expensive) telescope will have a better (theoretical) limit of angular resolution than one only 12 cm in diameter, in practice its resolution is limited by atmospheric seeing. The main reason that very large ground-based telescopes are built is to increase the available light-gathering power.

QUESTION 2.4

Some *advantages* of reflecting telescopes over refractors are:

(i) larger apertures are possible and hence higher light-gathering power and better angular resolution are achievable,

(ii) a Cassegrain telescope can have a relatively long focal length within a short tube and hence higher angular magnification can be achieved,

(iii) reflectors are easier to manufacture,

(iv) reflectors are not subject to chromatic aberration and we may reduce spherical aberration by using a paraboloidal mirror or a Schmidt correcting plate.

Some *disadvantages* of reflecting telescopes are: (i) they have higher losses of intensity through absorption, (ii) there is a gradual deterioration of the reflecting surfaces with age.

QUESTION 2.5

(a) The Schmidt correcting plate is a transparent glass plate built into the 'mouth' of a reflecting telescope. Its shape is calculated so as to introduce some differential refraction to various parts of the incoming wavefront, in such a way as to compensate for the spherical aberration of the main mirror. The result is good resolution over a significantly wider field-of-view.

(b) The important point to note for a Schmidt–Cassegrain telescope is that the light passes through the Schmidt plate first, and then reflects off the primary (objective) mirror. The remainder of the telescope is the same as the normal Cassegrain type. The overall design is shown in Figure 2.14.

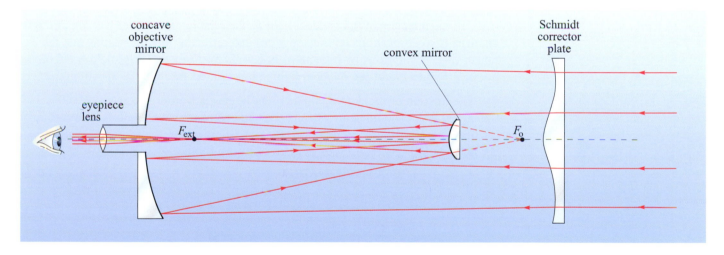

Figure 2.14 A Schmidt–Cassegrain telescope as discussed in Question 2.5.

QUESTION 2.6

(a) The image scale is determined by just one value: the focal length, f, of the telescope. The f-number, 10, is the focal length divided by the diameter, i.e. $f/300$ mm = 10, so the focal length must be $f = 3000$ mm. The image scale is then $I/(\text{arcsec mm}^{-1}) = 206\,265/(f/\text{mm}) = 206\,265/3000 = 69$ (to two significant figures), i.e. the image scale is 69 arcsec mm^{-1}.

(b) If 69 arcsec covers 1 mm, then 14 arcsec will cover 14 arcsec/69 arcsec mm^{-1} = 0.20 mm. Similarly, 25 arcsec will cover 25 arcsec/69 arcsec mm^{-1} = 0.36 mm.

QUESTION 3.1

We use the grating equation: $d(\sin \alpha + \sin \beta) = n\lambda$, but note that since the light strikes the grating at normal incidence, $\alpha = 0°$. The groove spacing, $d = 1/(120 \text{ mm}^{-1}) = 8.33 \times 10^{-6}$ m. The angle of diffraction of the light in successive orders is given by $\sin \beta = n\lambda/d$. So for red light the angles are given by $\sin \beta_r = n_r \times 700 \times 10^{-9}$ m$/8.33 \times 10^{-6}$ m and for blue light the angles are given by $\sin \beta_b = n_b \times 400 \times 10^{-9}$ m$/8.33 \times 10^{-6}$ m. These diffraction angles are listed in Table 3.1. It can be seen that the blue end of the third-order spectrum will overlap with the red end of the second-order spectrum. The overlap then becomes even more severe at higher orders.

Table 3.1 The diffraction angles calculated in Question 3.1.

red light		blue light	
n_r	β_r	n_b	β_b
1	4.8°	1	2.8°
2	9.7°	2	5.5°
3	14.6°	3	8.3°
4	19.6°	4	11.1°
5	24.8°	5	13.9°

QUESTION 3.2

(a) Call the intensity of the incident light I_0, then the intensity after passing through one lens (two absorptions on transmission) and reflecting off the grating and the mirror (two absorptions by reflection) is $I_0 \times 0.85 \times 0.85 \times 0.96 \times 0.96 = 0.67I_0$. So, only two-thirds of the incident light is transmitted through the spectrograph.

(b) The fraction of incident light emerging in the first-order spectrum is $0.40 \times 0.67I_0 = 0.27I_0$. The intensity of light per nanometre of wavelength is therefore $0.27I_0/300 = 9 \times 10^{-4}I_0$. So, the intensity per nanometre of wavelength, relative to the incident light, is reduced by a factor of over a thousand.

QUESTION 4.1

(a) The image scale is given by Equation 2.4 as

$$I /(\text{arcsec mm}^{-1}) = \frac{1}{4000 \times \tan(1 \, \text{arcsec})} = 51.6$$

(b) The field-of-view of the telescope on the detector is 10.0 mm/4000 mm = 2.5×10^{-3} radians, or $(2.5 \times 10^{-3} \times 180/\pi)$ degrees = 0.143°. This is equal to about 8.6′.

(c) 800 pixels correspond to a linear dimension of 10.0 mm, so the CCD has $800/10.0 = 80.0$ pixels mm^{-1}. The angular scale of the image is therefore 51.6 arcsec mm$^{-1}/80.0$ pixels mm^{-1} = 0.645 arcsec pixel^{-1}. Alternatively, the angular scale may be calculated as 8.6 arcminutes/800 pixels = 0.010 75 arcminutes pixel^{-1} which is identical to 0.645 arcsec pixel^{-1}.

(d) The stars are separated in the vertical direction. Their relative positions are given by the separations on the CCD of the two spectra. In this enlarged image, the centres of these two spectra are 50 mm apart and the whole frame measures 100 mm from top to bottom. Hence the two stars are $(50/100) \times 800 = 400$ pixels apart; this corresponds to an angular separation of (400 pixels $\times$ 0.645 arcsecond pixel^{-1}) = 258″ = 4.3′.

(e) The two emission lines labelled H$_\alpha$ and H$_\beta$ are about 53 mm apart and the image measures 100 mm from side to side. Hence, the two lines are $(53/100) \times 800 = 424$ pixels apart. They must therefore be separated in wavelength by (424 pixels $\times$ 0.4 nm pixel^{-1}) = 170 nm.

Note: In fact these are the first two lines in the Balmer series of the hydrogen spectrum and have wavelengths of 656.3 nm and 486.1 nm respectively.

QUESTION 5.1

The star on the left is saturated. This is apparent because the cross-section through the star is flat-topped. In the flat-topped regions, the pixels have accumulated their maximum value and cannot record a value that is proportional to the number of photons falling upon them. The problem may be avoided by taking several shorter exposures and averaging them together.

QUESTION 5.2

Flat fields correct for features such as dust particles on the filters which will vary from one filter to the next. The pixel sensitivity will also differ from one wavelength to another. Bias frames and dark frames are obtained with the shutter closed, so no light passes through a filter to reach the CCD. The bias level and dark current of the CCD depend on the electronics of its construction, not the light path.

QUESTION 6.1

We have already calculated the absolute visual magnitude of Rigel as $M_{\mathrm{Rigel}} = -7.12$. Using the same approach, the absolute visual magnitude of Ross 154 is $M_{\mathrm{Ross154}} = 10.45 + 5 - (5\log_{10}2.9) + 0 = 13.14$.

So using Equation 6.4, the ratio of luminosities in the visual band can be calculated as
$$10^{(M_{\mathrm{Rigel}} - M_{\mathrm{Ross154}})/2.5} = L_{\mathrm{Ross154}} / L_{\mathrm{Rigel}}$$

so $\quad L_{\mathrm{Ross154}}/L_{\mathrm{Rigel}} = 10^{(-7.12-13.14)/2.5} = 10^{-8.10} = 7.9 \times 10^{-9}$

and the ratio of the V-band luminosity of Rigel to that of Ross 154 is therefore $L_{\mathrm{Rigel}}/L_{\mathrm{Ross154}} = 1/7.9 \times 10^{-9} = 1.3 \times 10^8$. Therefore Rigel is about 130 million times more luminous than Ross 154 in the V-band.

QUESTION 6.2

The area of a circular aperture is πr^2 where r is the radius of the aperture. Hence the area of the source aperture is $\pi \times 12.0^2 = 452$ pixels. Similarly, the area of the annular background sky aperture is $(\pi \times 24.0^2) - (\pi \times 18.0^2) = 792$ pixels. The flux from the background sky is therefore $(7.00 \times 10^5/792) = 884$ photons per pixel. The background sky contribution to the flux in the star aperture is therefore (884 photons per pixel $\times$ 452 pixels) = 4.00×10^5 photons. The flux from the star itself in the aperture is therefore $(2.50 \times 10^6 - 4.00 \times 10^5) = 2.10 \times 10^6$ photons.

Now, using Equation 6.1 to find the instrumental magnitude of the star we have:
$$m_1 - m_2 = 2.5\log_{10}(F_2/F_1)$$
$$m_1 - 50.0 = 2.5\log_{10}(884/2.10 \times 10^6)$$
$$m_1 = 41.6$$

In other words, the star is 8.4 magnitudes brighter than the sky background in one pixel.

QUESTION 6.3

(a) A graph of (instrumental magnitude – catalogue magnitude) versus airmass for Regulus is shown in Figure 6.4. Two lines are shown on the graph: one for the V-band and one for the B-band. The gradients are measured to be $\varepsilon_V = 0.16$ magnitudes per airmass for V and $\varepsilon_B = 0.30$ magnitudes per airmass for B. The intercepts on the vertical axis are at $\zeta_V = 22.65$ magnitudes for V and $\zeta_B = 21.26$ magnitudes for B.

(b) The target star has an airmass of $1/\cos(90° - 35°) = 1.74$. Equation 6.6 is $(m' - m) = \varepsilon X + \zeta$. So in the V-band, $(m'_V - m_V) = (0.16 \times 1.74) + 22.65 = 22.93$. The V-band catalogue magnitude of the target star is therefore, $m_V = m'_V - 22.93 = 29.75 - 22.93 = 6.82$. Similarly in the B-band, $(m'_B - m_B) = (0.30 \times 1.74) + 21.26 = 21.78$. The B-band catalogue magnitude of the target star is therefore, $m_B = m'_B - 21.78 = 31.25 - 21.78 = 9.47$.

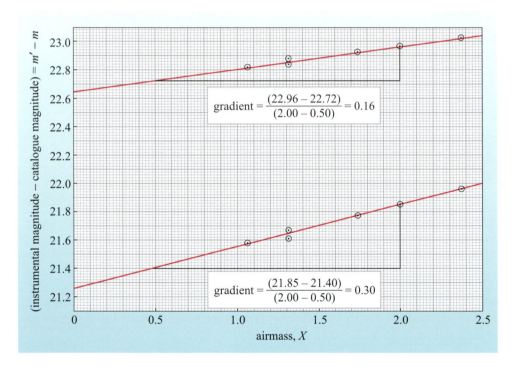

Figure 6.4   A graph of (instrumental magnitude – catalogue magnitude) versus airmass, in both the V- and B-bands, for the standard star Regulus in Question 6.3.

QUESTION 7.1

The horizontal stripes in Figure 3.9 are each segments of a single spectrum produced by an echelle spectrometer. Each stripe comprises a very small wavelength range of the spectrum of the object being observed. The vertical stripes in Figure 7.2a are emission lines in the spectrum of a copper–argon arc lamp. Because the light from the arc lamp fills the whole slit, the spectrum is extended in the spatial (vertical) direction.

QUESTION 8.1

As the resolution is directly proportional to the wavelength of the light, using monochromatic light at the blue end of the spectrum would offer the best resolution. It should be noted that other useful information would be lost in such conditions.

QUESTION 8.2

A sample answer follows. Yours will certainly not be identical to this, but you should be able to capture a similar summary in your own words.

The first polarizing filter allows *plane-polarized light* to fall on the sample. On entering the sample, the light is split into two plane-polarized beams with mutually perpendicular planes of polarization. The two beams experience different refractive indices within the sample, and so travel through it at different speeds. On emerging from the sample, the two beams will be out of phase with each other – the phase difference being a function of the wavelength of light, for a given mineral type and sample thickness. By passing the two beams through a second polarizing filter, the components of each beam that emerge have the same plane of polarization as each other and may be combined using a lens. The pattern of interference colours produced by the two combined beams depends on the value of the birefringence of the mineral under study. So by carefully studying the colours in the resulting image and comparing them to a calibrated chart, the minerals in the sample may be identified.

QUESTION 9.1

(a) You should recognize several impact craters in Figure 9.3 (outside the box), by virtue of their circular shape (when seen from above, as in this image, rather than obliquely). The largest crater appears very similar to the large example in Figure 9.2. Having recognized craters, you should be able to use the shadows to tell you that the sunlight is coming from the right.

(b) The long feature is brightly lit on its right-hand side and shadowed on its left-hand side. Given that the illumination is from the right, this must be a ridge. A valley would have shadow on the right (i.e., on the wall facing away from the Sun) and be brightly lit on the opposite, Sun-facing, wall.

QUESTION 13.1

(a) The mean value $\langle m_v \rangle$ is found by adding all of the magnitude values and dividing by the number of values (ten). Thus

$$\langle m_v \rangle = (220.0)/10 = 22.0$$

(b) The measurements are spread over a range from 21.6 to 22.3, a range of about ± 0.35 magnitudes. It is conventional to quote the uncertainty in a measured value as about 2/3 of the spread, in recognition of the fact that values from the extremes of the range are not very likely. So in this case we would estimate the uncertainty as $2/3 \times (\pm 0.35$ magnitudes), which is ± 0.2 magnitudes, to one significant figure.

(c) The standard deviation is calculated using the procedure in Box 13.1. The mean value was calculated in part (a). The sum of the squared deviations of the measurements from the mean is 0.40 magnitudes^2. The mean of the squared deviations is therefore (0.40 magnitudes^2)/10 = 0.04 magnitudes^2, and the standard deviation (calculated by taking the square root of this) is 0.2 magnitudes. Note that the size of the uncertainty estimated in (b) is approximately the same as the standard deviation, and this is why it is often sufficient to use the simpler 2/3 spread procedure.

(d) The uncertainty in the mean magnitude is $\sigma_m = s_n/\sqrt{n} = 0.2/\sqrt{10} \sim 0.06$ magnitudes.

QUESTION 13.2

(a) The uncertainty σ_m in the mean value of n measurements is related to the standard deviation s_n of the measurements by $\sigma_m = s_n/\sqrt{n}$. So $\sigma_m = 6$ nm$/\sqrt{10} \sim 2$ nm. Note that this is much smaller than the uncertainty in a single measurement, which is represented by the standard deviation of 6 nm.

(b) The difference between the mean value (585 nm) and the suspected true value (591 nm) is 6 nm, which is three times larger than the uncertainty in the mean. Assuming that values of the means that would be obtained from many sets of ten measurements have a Gaussian distribution, then the probability of the value of the mean differing from the true value by three times the uncertainty in the mean is only 0.003. It is therefore unlikely, though possible, that the true value is 591 nm.

(c) If $\sigma_m = 1$ nm, and $s_n = 6$ nm, then $\sqrt{n} = s_n/\sigma_m = 6$, and so $n = 36$. So reducing the uncertainty from 2 nm to 1 nm would require almost four times as many measurements.

QUESTION 13.3

The noise from the star and the background is given by

$$\text{noise} = \sqrt{N_{star} + (N_{back} \times np)} = \sqrt{2.5 \times 10^6 + 7.00 \times 10^5} = 1790$$

The uncertainty due to counting statistics in the star's astronomical magnitude is

$$\delta m = 2.5 \log_{10}\left(1 + \frac{\delta F}{F}\right)$$

$$= 2.5 \log_{10}\left(1 + \frac{1790}{2.50 \times 10^6}\right)$$

$$\approx 0.001 \text{ magnitudes}$$

QUESTION 15.1

The graph contains several examples of bad practice:

- The airmass has been plotted up the vertical axis and the difference between the instrumental and catalogue magnitudes along the horizontal axis. Since the magnitude difference is the dependent variable (i.e. it is the measured quantity), the axes should be the other way round.

- The graph does not have an informative title.

- The scale that has been chosen – 7 squares per magnitude – does not make it easy to plot the points accurately!

- Each of the points is plotted with a large cross whose centre is difficult to determine.

- None of the points has an uncertainty bar.

- The straight line drawn is clearly not the best fit to the data.

- There is an uneven distribution of points with airmass. Ideally, some more standard stars should have been observed at intermediate values of airmass.

QUESTION 15.2

Kepler's third law can be written $a^3 = P^2 \times (GM/4\pi^2)$. Therefore a graph of a^3 versus P^2 should yield a straight line, passing through the origin (since the equation predicts that $a = 0$ when $P = 0$). The gradient of the line will be equal to $(GM/4\pi^2)$. A variant of this would be to plot a versus $P^{2/3}$ or $a^{3/2}$ versus P, but these are really the same sort of graph as that just described.

Alternatively, taking logs of both sides of Kepler's third law, we obtain

$$3\log(a) - 2\log(P) = \log(GM/4\pi^2),$$

or $\qquad \log(a) = (2/3)\log(P) + (1/3)\log(GM/4\pi^2)$

Therefore a graph of $\log(a)$ versus $\log(P)$ will yield a straight line whose gradient is 2/3 and whose intercept is at $(1/3)\log(GM/4\pi^2)$.

REFERENCES

Filippenko, A. and Greenstein, J. L., '*Faint spectrophotometric standard stars for large optical telescopes*', 1984, Publications of the Astronomical Society of the Pacific, **96**, pp. 530–536

Landolt, Arlo U., '*UBVRI photometric standard stars in the magnitude range 11.5–16.0 around the celestial equator*', 1992, Astronomical Journal,. **104**, 1, pp. 340–371 & pp. 436–491.

Oke, J. B. and Gunn, J. E., '*Secondary standard stars for absolute spectrophotometry*', 1983, Astrophysical Journal, Part 1, **266**, pp. 713–717.

WEBSITES

Digitized sky survey	*http://ledas-www.star.le.ac.uk/DSSimage/*
Digitized sky survey	*http://stdatu.stsci.edu/dss/*
Solar system ephemeris generator	*http://ssd.jpl.nasa.gov/cgi-bin/eph*
Centre de Données astronomique	*http://cdsweb.u-strasbg.fr/*
NASA/IPAC Extragalactic Database	*http://nedwww.ipac.caltech.edu/*
NASA Astrophysics Data System	*http://adswww.harvard.edu/*
NASA Astrophysics Data System	*http://ukads.nottingham.ac.uk/*
Astro-ph E-print archive	*http://uk.arxiv.org/archive/astro-ph*
General archives of NASA mission images:	*http://photojournal.jpl.nasa.gov*
	http://wwwflag.wr.usgs.gov/USGSFlag/Space/wall/wall.html
	http://pds.jpl.nasa.gov/planets/
The Nine Planets (multimedia tour of the Solar System):	*http://seds.lpl.arizona.edu/billa/tnp/*

These urls were all correct at the time of going to press.

ACKNOWLEDGEMENTS

Grateful acknowledgement is made to the following sources for permission to reproduce material within this product.

Cover

Background image: © 1979 Anglo–Australian Observatory, photograph by David Malin;

Thumbnail image on extreme left: © Anglo–Australian Observatory.

Figures

Figure 0.1: © Gordon Garrard; *Figure 0.3a*: © NASA; *Figure 0.3b*: courtesy of the James Clerk Maxwell Telescope, Mauna Kea Observatory, Hawaii; *Figure 0.3c*: © NRAO/AUI/NSF; *Figure 0.3d*: © ESA; *Figure 0.3e*: © NASA; *Figure 1.5*: T. Credner, AlltheSky.com; *Figure 1.10*: © 1979 Anglo–Australian Observatory, photograph by David Malin; *Figure 2.4*: © UCO/Lick Observatory; *Figure 2.10a*: courtesy of Gemini Observatory; *Figure 3.10a&b*: © Anglo–Australian Observatory; *Figure 4.1*: John Walsh/Science Photo Library; *Figure 9.1*: © NASA; *Figure 9.2*: © NASA; *Figure 9.3*: © NASA.

Every effort has been made to contact copyright holders. If any have been inadvertently overlooked the publishers will be pleased to make the necessary arrangements at the first opportunity.

GLOSSARY

The glossary includes all **bold** terms from the book, plus a number of other key words that are relevant to practical observations in astronomy and planetary science. Cross-references within the glossary are indicated by *italicized* words.

absolute magnitude (*M*)   The *apparent magnitude m* of a star which would be obtained at a standard distance of 10 *parsec* from the star, in the absence of *interstellar extinction*. Absolute magnitudes may be quoted in any waveband, for example 'absolute B-band magnitude', but if no band is specified, then V-band, i.e. visual, is assumed. Absolute visual magnitudes of stars range from $M_V \sim -10$ for massive supergiants (such as Eta Carina) to $M_V \sim +15$ for low-mass main-sequence dwarfs (such as Proxima Centauri). This range of ~25 magnitudes corresponds to a range of $\sim 10^{10}$ in visual *luminosity*.

absorption   The decrease in intensity (of flux density) of radiation as it passes through a medium (e.g. the *interstellar medium* or the Earth's atmosphere) due to the photons being absorbed by atoms or molecules. See *interstellar extinction* and *atmospheric extinction*.

absorption lines   Narrow *wavelength* or *frequency* ranges in a *spectrum* where the *spectral flux density* is less than at adjacent wavelengths (or frequencies). Absorption lines are produced when light, or other *electromagnetic radiation*, passes through cooler material. Radiation is absorbed at certain specific wavelengths, which depend on the absorbing material, so that a pattern of dips and troughs is imposed on the *continuous spectrum*.

absorption spectrum   A spectrum in which *absorption lines* predominate.

accurate   Accurate measurements are those in which the *systematic uncertainty* is small.

adaptive optics   A technique to remove the effects of atmospheric turbulence by rapidly correcting the shape of the *primary mirror* or *secondary mirror* in a telescope. *Point spread functions* close to the *diffraction-limited* value of the telescope may be obtained.

airmass   The ratio of the thickness of the atmosphere at the observing *altitude* to the thickness at the *zenith*. It is approximated as airmass = 1/cos(*zenith angle*).

alt-azimuth mounting   A telescope with an alt-azimuth mounting may be moved independently in the horizontal (azimuth) and vertical (altitude) directions. Although simple to construct, such mountings require more complex movements when tracking astronomical objects than do telescopes with an *equatorial* mounting.

altitude   The angle between a horizontal plane and the direction to an astronomical object, measured along the *great circle* that passes through the *zenith* and the point on the *celestial sphere* in the direction to the object. It is equal to 90° minus the *zenith angle*.

analyser   A second *polarizing filter* used in a *petrographic microscope* to combine the two *plane-polarized light* beams produced by *birefringence* in a crystal.

angular dispersion   When considering a *diffraction grating*, this is a measure of how large a change in the diffraction angle (β) results from a specified change in wavelength (λ). It may be derived from the *grating equation* as $\partial\beta/\partial\lambda = (1/\cos\beta) \times (n/d)$ where n is the *spectral order* of the diffraction pattern and d is the *groove spacing* of the grating.

angular magnification   The angle subtended by the image of an object seen through a telescope divided by the angle subtended by the object without the aid of a telescope. Equal to f_o/f_e where f_o is the focal length of the primary or objective lens (or mirror) and f_e is the focal length of the eyepiece lens.

aperture   A (usually) circular zone on a CCD frame surrounding the image of an object of interest, within which the flux of light is measured.

aperture photometry   *Photometry* performed on objects within a *CCD* image by measuring the flux within a given *aperture*.

apparent magnitude (*m*)   A measure of the apparent brightness of a body. For a star, it is a measure of the flux density received. Apparent magnitudes may be quoted in any waveband, for example 'apparent B-band magnitude', but if no band is specified, then V-band, i.e. visual, is assumed. The brightest star in the sky (Sirius) has an apparent visual magnitude $m_V \sim -1.5$, while the faintest stars visible to the naked eye are 1000 times fainter with apparent visual magnitudes $m_V \sim +6$. See *absolute magnitude*.

arc lamp   In astronomy, a hollow cathode lamp whose spectrum is used to calibrate the *wavelength* of astronomical spectra.

astronomer   In the context of this book, any scientist engaged in research in any field of astrophysics, cosmology or planetary science.

astronomical unit (AU)   The average distance from the Earth to the Sun. More precisely, the semimajor axis of the

Earth's elliptical orbit around the Sun. (In fact, the Earth-Sun distance varies very slightly, so the astronomical unit is now defined very exactly as a fixed quantity.) Equal to about 1.50×10^8 km.

atmospheric absorption   *Absorption* of light from astronomical objects that occurs in the Earth's atmosphere. One component of *atmospheric extinction*.

atmospheric extinction   The combination of *absorption* and *scattering* of light from astronomical objects that occurs in the Earth's atmosphere. It may be quantified by an *extinction coefficient*.

atom   The smallest characteristic entity of a *chemical element*. An atom consists of a positively charged *nucleus* surrounded by a cloud of one or more negatively charged *electrons*.

atomic number   The number of *protons* within the *nucleus* of an *atom*. It is also equal to the number of *electrons* in the neutral atom. Each *chemical element* has a unique, characteristic atomic number.

bad pixel   A pixel on a *CCD* that is faulty and returns a value that is misrepresentative of the light falling on it.

bell curve   See *Gaussian distribution*.

bias   When the signals from a *CCD* are digitized, an offset known as the bias is intentionally introduced into the digital value. The bias value must subsequently be removed when analysing the CCD images.

bias frames   Entire images created by reading out a *CCD* following a zero-second exposure. Such an image contains no *photoelectrons*, so the intensity value across the image is simply the *bias* value at each point. See *overscan strips*.

birefringence   The phenomenon by which a beam of *plane-polarized light* is split into two refracted beams when it enters certain types of crystal. Both beams are also plane polarized, but at 90° to each other. The *refractive index* of the crystal is different for each of the two refracted beams. The birefringence is quantified by the difference between the two refractive indices.

brightness   In astronomy, this term is used rather loosely to refer to the amount of light, or other radiation, received from or emitted by an astronomical body. See also *flux*, *luminosity*, *apparent magnitude*, *absolute magnitude*.

broad-band photometric wavebands   A set of wavelength ranges commonly used in observations of astronomical objects. The wavebands are: U, ultraviolet, central wavelength 365 nm; B, blue, 440 nm; V, visual (i.e. green–yellow), 550 nm; R, red, 700 nm; and I, near infrared, 900 nm. This classification is known as the *Johnson photometric system*. There are also three commonly used bands in the infrared known as J, H and K with central wavelengths at 1.25 μm, 1.65 μm and 2.2 μm respectively.

camera (lens or mirror)   In a *spectrograph*, the lens or mirror used to focus the light emerging from the *prism* or *diffraction grating* onto the detector.

Cassegrain telescope   A *reflecting telescope* incorporating a concave *primary mirror* and a convex *secondary mirror* to extend the *effective focal length* of the instrument.

CCD (charge-coupled device)   A photoelectric detector, i.e. one that responds to incoming photons by producing an electrical signal. CCDs comprise a set of *pixels* arranged in a two-dimensional array that may be a few thousand elements along each side. Physically, CCDs are typically no more than a couple of centimetres in size.

celestial equator   The projection of the Earth's Equator from the centre of the Earth onto the *celestial sphere*.

celestial sphere   An imaginary sphere with the same centre as the Earth on which the positions of astronomical objects may be mapped. The radius of the celestial sphere is assumed to be very large, but is otherwise not specified.

chemical element   A substance that consists of only one type of *atom*. For example, the chemical element copper is made only of copper atoms. A chemical element cannot be broken down into simpler constituents by using chemical reactions. All the atoms in a sample of a chemical element have the same *atomic number*, but may have different *mass numbers* (see *isotope*).

chromatic aberration   An aberration inherent in all lenses that causes light of different *wavelengths* to be brought to a focus at different points. It arises due to the fact that the *refractive index* is usually a function of wavelength. It may be corrected using compound lenses comprising lenses of different strengths and different materials.

civil time   The conventional 24 hours per day used for normal timekeeping. The civil time is defined to be the same within each time zone of the Earth, although the solar time (determined from the position of the Sun) will vary across each zone.

coherent (source of light)   A spatially extended source of waves is said to be coherent if the emissions contributed by the various regions of the source are always in phase with one another.

collimated   Used to describe a beam of light that has been rendered parallel, often by a *collimator* lens or mirror.

collimator   A lens or mirror in a *spectrograph* used to render the incoming light parallel before it encounters the *prism* or *diffraction grating*.

colour index   The difference, measured in *magnitudes*, between the brightness of an object at one *wavelength* and its brightness at another, different wavelength. For example, using measurements in the blue 'B' and visual 'V' *broad-band photometric wavebands*, the colour index is denoted $m_B - m_V$ or simply $B - V$. In certain cases, the colour index can indicate the *temperature* of an astronomical object.

constellation   One of 88 irregularly shaped, but contiguous, regions of the sky based around patterns of stars as seen from Earth.

constructive interference   The result of a *superposition* of two or more similar waves that are in phase at a given point.

continuous spectrum   A *spectrum* that is broad and smooth, i.e. the spectral flux density exhibits no sharp changes with *wavelength*.

continuum normalized spectrum   A *spectrum* that has been divided pointwise by a smooth curve fit to its continuum level. The resulting normalized spectrum has a mean value of one, with *emission lines* and *absorption lines* superimposed on top.

coordinated universal time (UTC)   A more accurate form of *universal time* based on atomic time. To correct for the irregular varying motion of the Earth around its axis, and in its orbit around the Sun, UTC is kept in step with UT by the insertion or deletion of *leap seconds* at the end of June or December as necessary.

cosmic ray event   An image *pixel* on a *CCD* frame that registers an anomalously high intensity value, caused by a cosmic ray striking the CCD during the exposure.

cross-polars   When a sample is viewed in a *petrographic microscope* using an initial *polarizing filter*, and a second polarizing filter known as the *analyser*, it is said to be viewed in cross-polars. The resulting coloured *interference* pattern can help to identify the minerals under study.

dark current   A signal generated in a *CCD* even when no light is falling on it. It can be minimized by cooling the CCD.

daylight saving time   A period in which *civil time* is advanced by an hour in the spring and summer months.

December solstice   The time at which the South Pole of the Earth is maximally tilted towards the Sun. It is known as the winter solstice in the Northern Hemisphere and the summer solstice in the Southern Hemisphere. It occurs around 21 December.

declination   The celestial equivalent of latitude, abbreviated as *dec* or δ, and measured in degrees, arcminutes, and arcseconds. It extends from 0° at the *celestial equator* to +90° at the *north celestial pole* and −90° at the *south celestial pole*.

declination axis   In an *equatorial* mounting for a telescope, this is the axis at right angles to the *polar axis*.

destructive interference   The result of a *superposition* of two or more similar waves that are exactly out of phase with one another (i.e. have a phase difference of π radians) at a given point.

differential photometry   *Aperture photometry* of a target with respect to the flux from another object on the same *CCD* image. The difference in *magnitude* is recorded, but it is not necessary to know the magnitude of the reference, or comparison object. See also *relative photometry*.

diffraction   When a wavefront meets a partial obstacle or aperture, the waves spread out and this gives waves the ability to 'bend round corners'. The phenomenon is known as diffraction.

diffraction grating   A device used to disperse light in a *spectrograph*. In the commonly used *reflection grating* (or *reflective diffraction grating*), the surface of the device has many parallel grooves, or steps, cut into it. The angles at which light is diffracted by the grating are given by the *grating equation*.

diffraction-limited   If a telescope produces images such that the *theoretical limit of angular resolution* is achieved, it is said to be operating under diffraction-limited conditions.

diffraction pattern   This is a term for the pattern of *interference fringes* produced by the *superposition* of waves that have been diffracted by an object such as a *diffraction grating*. Maxima and minima are caused by *constructive interference* and *destructive interference* respectively.

digitized sky survey   A computerized map of the whole sky produced by scanning the photographic plates obtained by the Oschin Schmidt Telescope at Palomar Observatory and the UK Schmidt Telescope at Siding Spring Observatory.

dispersion   The process of making light of different *wavelengths* travel in different directions. It may be achieved using either a *prism* or a *diffraction grating* in a *spectrograph*.

distance modulus   The difference between the *apparent magnitude* (m) of an astronomical object and its *absolute magnitude* (M). Given by

$$m - M = 5 \log_{10}(d/\mathrm{pc}) - 5 + A$$

where *d* is the distance to the object (in *parsec*) and *A* is the amount of *interstellar extinction* (in *magnitudes*) along the line of sight.

dome flats *Flat fields* obtained by exposing a *CCD* to the uniformly illuminated inside of an observatory dome. See *sky flats*.

double refraction See *birefringence*.

echelle spectrograph An echelle spectrograph has a second dispersing element, either a second *grating* or a *prism*, which disperses the light at right angles to the direction of dispersion produced by the main *grating*. The effect is to produce a spectral image that consists of a stacked series of spectra. Each of the stacked spectra represents a part of the *spectrum* of the object, spanning only a very narrow range of *wavelength*.

ecliptic The complete trace of the Sun's coordinates on the *celestial sphere* over the course of a year.

ecliptic plane The plane of the Earth's orbit, extended out to meet the *celestial sphere*.

effective focal length (of a reflecting telescope) This is the focal length of a single mirror having the same diameter as the *primary mirror* of a telescope that would give a cone of light that converges at the focus at the same angle as the two-mirror system.

electrical energy *Potential energy* that an object has because it is electrically charged and separated from other electrically charged objects.

electric charge A fundamental property of matter. There are two types, known as positive and negative charge. Like charges repel each other, and unlike charges attract each other. Objects with no charge, or with equal amounts of positive and negative charge, are electrically neutral.

electromagnetic radiation Radiation comprising any part of the *electromagnetic spectrum*.

electromagnetic spectrum A collective term used to describe the various *wavelength* ranges of electromagnetic radiation. In order of increasing wavelength, these ranges are *γ-ray*, *X-ray*, *ultraviolet*, *visible light*, *infrared*, *microwave* and *radio wave*. See also *spectrum*.

electromagnetic wave A fluctuating pattern of electric and magnetic fields, in which each field takes the form of a wave. At any point in an electromagnetic wave, the electric and magnetic fields are mutually perpendicular, and each field is also perpendicular to the direction of propagation of the wave. The waves travel through a vacuum at the *speed of light*. Electromagnetic waves of appropriate *wavelength* (or frequency) may be used to model each of the kinds of *electromagnetic radiation* that comprise the *electromagnetic spectrum*.

electron (e⁻) A fundamental particle with negative *electric charge*.

electronvolt (eV) A unit of *energy* equivalent to the energy gained by an electron in moving through a potential difference of one volt. *Photons* of visible light have an energy of around 2 to 3 eV, where $1 \text{ eV} = 1.602 \times 10^{-19}$ J.

emission lines Narrow *wavelength* or *frequency* ranges in a *spectrum* where the *spectral flux density* is greater than at adjacent wavelengths (or frequencies). Emission lines are produced when atoms make transitions from higher to lower *energy levels*.

emission spectrum A spectrum in which *emission lines* predominate.

energy A physical property possessed by an object that measures the capacity of the object to make changes to other objects. There is a variety of possible changes, and these include changes in position and changes in motion. Energy has various forms, including *kinetic energy*, *gravitational energy* and *electrical energy*, but the *law of conservation of energy* applies in all processes that involve conversions or transfers of energy. The SI unit of energy is the *joule*. Energy may also be quantified using the unit of *electronvolt* (eV).

energy-level diagram A pictorial way of representing *energy levels*. Energy levels corresponding to low energy are drawn at the bottom of the diagram.

energy levels Specific values of energy that an atom is allowed to have. *Transitions* occur between energy levels, in the process of which *photons* are emitted or absorbed.

equation of a straight line An equation of the form $y = mx + c$, where *m* and *c* are constants that respectively represent the *gradient* and the *intercept* of the line on the *y*-axis.

equatorial mounting A telescope with an equatorial mounting has its *polar axis* aligned with the Earth's rotational axis, and its *declination axis* at right angles to this. Although more complex to construct than *alt-azimuth* mountings, equatorial mountings allow astronomical objects to be tracked by simply rotating the telescope around the polar axis.

equinox A date appended to celestial coordinates to specify the year at which they are correct (see *precession*). Note also *March equinox* and *September equinox* which apply the word in a different sense.

errors See *uncertainties*.

error bars See *uncertainty bars*.

extinction   See *atmospheric extinction* and *interstellar extinction*. Generally refers to a combination of *absorption* and *scattering* of light.

extinction coefficient (*ε*)   Characterizes the extinction (i.e. *absorption* and *scattering*) by the Earth's atmosphere as a function of *wavelength*. In order to calibrate photometric observations, the extinction coefficient must be determined in each waveband used. The instrumental magnitude (*m′*) and catalogue magnitude (*m*) of an astronomical object are related by $(m' - m) = \varepsilon X + \zeta$ where *ε* (the Greek letter epsilon) is the extinction coefficient in magnitudes per airmass, *X* is the *airmass* and *ζ* (the Greek letter zeta) is the *zero-point offset* between instrumental and apparent magnitudes.

eyepiece lens   The lens in a telescope or microscope used to convert the *real image* produced by the *objective* into a *virtual image*, for observation by the eye.

f-number   The ratio of the *focal length* of an optical instrument to the diameter of the bundle of parallel light rays that is brought to a focus (i.e. often the diameter of the main optical element).

fibre-fed spectrograph   A *spectrograph* in which optical fibres are used to obtain the spectra of multiple point-like objects simultaneously.

field-of-view (of a telescope)   The area of sky that is accessible at one time, expressed in terms of an angular diameter. The angular field-of-view, in radians, is equal to the diameter of *field stop* in the *eyepiece* or the linear size of the detector, divided by the effective *focal length* of the *primary* mirror or lens.

field stop   An aperture built into an *eyepiece* which determines the angular *field-of-view* of an optical instrument such as a *telescope*.

finding chart   An image of a small patch of sky (typically a few arcminutes across) centred on the object of interest. It is used to help locate and identify the target when pointing the telescope at an object.

First Point of Aries (♈)   The point on the *celestial equator* at which the Sun's *declination* changes from south to north. It defines the zero line of *right ascension*. The Sun passes through the First Point of Aries at the *March equinox*.

flat field   An image obtained by exposing a *CCD* to a uniformly illuminated light source. Flat fields are used to calibrate for the variation in *pixel*-to-pixel sensitivity of a CCD. See *dome flats* and *sky flats*.

flux   A measure of the number of photons per second from an astronomical object received by a detector. It may be represented by the *apparent magnitude* of the object.

focal length   The distance of the *focal point* from the centre of a lens or mirror.

focal point   The point on the optical axis of a convex lens or concave mirror to which parallel rays of light converge. Alternatively, the point on the optical axis of a concave lens or convex mirror from which parallel rays of light appear to diverge.

frame   An alternative name for a *CCD* image.

frequency   The number of *wavelengths* in a wave that pass a fixed point in space in unit time.

fringes   Wave-like patterns of intensity variation across a *CCD* image. See *fringing*.

fringing   An effect caused by multiple reflections, within a *CCD* or filters, of light at a single *wavelength*. It gives rise to *fringes* on CCD frames but may be corrected using *sky flats*.

galaxy   An ensemble of *stars*, gas and dust. Our own Galaxy, the Milky Way, is a typical spiral galaxy with a mass about 10^{11} times that of the *Sun*.

Galilean telescope   A *refracting telescope* consisting of a convex objective *lens* and a concave *eyepiece* lens.

γ-ray   The region of the *electromagnetic spectrum* corresponding to the very shortest *wavelengths*, or highest energies.

gamma-ray   See *γ-ray*.

gas   A fluid phase of matter characterized by the lack of a definite volume or shape, other than that imposed by a container.

Gaussian distribution   Also known as a bell curve or normal distribution, this function applies to many populations and indicates how a variable property is distributed about a *mean* (or average) value. The values may be different measurements of a single quantity, in which case the spread reflects the size of the measurement *uncertainties*. Alternatively, the values may genuinely differ from one object to another in a population, e.g. height, in which case the spread reflects the variety in the population. The spreads of many populations in nature resemble Gaussian distributions. The equation for a Gaussian distribution of measurements of *x* is

probability of finding a value $x \equiv$

$$G(x) = \frac{1}{s_n \sqrt{2\pi}} \exp\left(\frac{-(x - \langle x \rangle)^2}{2s_n^2}\right)$$

where $\langle x \rangle$ is the mean and s_n is the *standard deviation* which quantifies the width (or spread) of the distribution. The Gaussian distribution peaks at the mean, and is symmetric about that peak. About 68% of values lie within $\pm s_n$ of the mean.

gradient (of a graph)   The slope of a straight-line graph. It is calculated as the 'rise' divided by the 'run' i.e. the change in vertical coordinate divided by the corresponding change in horizontal coordinate. See *equation of a straight line*.

graticules   Measuring scales and crosshairs fitted to *eyepieces* in microscopes or telescopes and used for positioning and measurement.

grating equation   The equation that describes the angles at which light is diffracted by a *diffraction grating*. Given by

$$d(\sin\alpha + \sin\beta) = n\lambda$$

where α is the angle between the incident light and the normal to the grating, β is the angle between the diffracted light and the normal to the grating, d is the spacing between adjacent grooves on the grating, n is the *spectral order*, and λ is the *wavelength* of the light in question.

gravitational energy   Work has to be done against the *gravitational force* to raise an object to a greater height, and the *energy* transferred is stored as gravitational energy. This energy is released and can do work when the object falls. Gravitational energy is a form of *potential energy*.

gravitational force   The force produced by an object which possesses *mass*.
(See also *Newton's law of gravitation*.)

great circle   Any line on a sphere that is the intersection between the sphere and a plane passing through its centre. Examples include the *meridian* and the *celestial equator*.

groove spacing d The distance between adjacent grooves on a *diffraction grating*.

image scale   When a *telescope* is used to form an image directly onto an electronic or photographic detector, the image scale is given by

$$I/\text{arcsec mm}^{-1} = [(f_o/\text{mm}) \times \tan(1\text{ arcsec})]^{-1}$$

Image scale indicates how many arcseconds on the sky are projected onto 1 mm in the image plane.

infrared   The region of the *electromagnetic spectrum* corresponding to *wavelengths* shorter than those of *microwaves* and longer than those of *visible light*.

instrumental magnitude   The *magnitude* measured on a *CCD* image relative to some fixed reference *brightness*, such as the brightness of the background sky.

integral field unit spectrograph   A device in which optical fibres feed light from an extended region of a telescope's image plane into a *spectrograph*, in order to produce multiple spectra from an extended object.

integrating detector   An astronomical detector that is able to build-up (integrate) the total light falling on it over an extended exposure.

intercept (of a graph)   The point on the vertical axis at which a straight line graph crosses the axis, i.e. the vertical coordinate corresponding to a horizontal coordinate of zero. See *equation of a straight line*.

interference   The result of the *superposition* of two or more waves derived from an extended, but *coherent* source. See also *constructive interference* and *destructive interference*.

interference fringes   These make up the pattern of light consisting of bright and dark regions caused by *constructive interference* and *destructive interference*, respectively.

interplanetary dust particles   Dust particles orbiting the Sun and produced mainly by the decay of comets.

interstellar absorption   *Absorption* of light from astronomical objects that occurs in the intervening gas and dust between the object and the Earth. One component of *interstellar extinction*.

interstellar extinction   The combination of *absorption* and *scattering* of light from astronomical objects that occurs in the intervening gas and dust between the object and the Earth. It may be quantified in terms of an equivalent number of *magnitudes*.

interstellar medium   The *gas* and dust lying between the stars which is responsible for partially absorbing and scattering the light emitted by them, and so causing stars to appear less bright than they would otherwise be.

ion   An *atom* or a group of atoms that has lost or gained one or more *electrons*, leaving it with a net positive or negative *electric charge*.

ionize   When one or more *electrons* are removed from an *atom*, the atom is said to be ionized.

ionization   The process by which an *atom* (or existing *ion*) is turned into an *ion* (or more positive *ion*) when one or more electrons are removed from it.

ionization energy   The *energy* required to completely remove one *electron* from an *atom* or existing *ion*.

isotope   Atoms with the same number of *protons* in their nuclei, but different numbers of *neutrons* are called isotopes. Because they have the same number of protons, they have the same *atomic number* and are atoms of the same *chemical element*. But because of the different number of neutrons, they differ in *mass number*.

Johnson photometric system   See *broad-band photometric wavebands*.

joule (J)   The SI unit of energy. $1\text{ J} = 1\text{ kg m}^2\text{ s}^{-2}$.

June solstice   The time at which the North Pole of the Earth is maximally tilted towards the Sun. It is known as the summer solstice in the Northern Hemisphere and the winter solstice in the Southern Hemisphere. It occurs around 21 June.

kelvin (K)   The SI unit of *temperature*. The absolute zero of temperature is at 0 K, water freezes at about 273 K, and water boils at about 373 K.

Keplerian telescope   A *refracting telescope* consisting of a convex *objective* lens and a convex *eyepiece* lens.

Kepler's third law   Where one body is in orbit around another, the square of the orbital period is proportional to the cube of the semimajor axis of the ellipse. It may be explained in terms of Newton's laws of motion and *Newton's law of gravitation*.

kinetic energy   *Energy* associated with motion of an object. Depends on the *mass* of the object and the speed at which it is moving.

law of conservation of energy   States that the total amount of *energy* in an isolated system is constant, since energy is neither created nor destroyed.

leap seconds   Intervals of time, inserted or deleted at the end of June or December as necessary, to correct for the irregular varying motion of the Earth around its axis, and in its orbit around the Sun. Their effect is to keep *coordinated universal time* and *universal time* in step.

least-squares method   A method of finding the best-fit formula to a set of data by minimizing the sum of the squares of the deviations of the data from the fitted formula.

lens   A piece of transparent material (such as glass, plastic or quartz) that is specially shaped to alter the path of a beam of light by *refraction*. A lens usually has two faces each being part of a sphere. A lens may cause a parallel beam of light either to diverge or to converge.

librations   Apparent oscillations of the Moon's angular position about its mean value, allowing us to see somewhat more than 50% of its surface from Earth, despite its *synchronous rotation*.

light curve   A diagram of the variation of *brightness* (e.g. magnitude, flux density) or *luminosity* with time, of a celestial object such as a variable *star* or rotating asteroid.

light-gathering power (of a telescope)   Given by $(D_o/D_p)^2$ where D_o is the diameter of the *primary mirror* (or lens) and D_p is the diameter of the eye's pupil.

light source   The tungsten–halogen bulb used in a petrographic microscope to illuminate the sample.

light-year (ly)   The distance that electromagnetic radiation would travel in a year at the *speed of light* in a vacuum. $1\text{ ly} = 9.46 \times 10^{15}$ m.

limit of angular resolution   The minimum angular separation at which two equally bright stars would be just distinguished by an astronomical telescope. For an instrument whose *primary* mirror (or *lens*) has a diameter D_o it is given by $\alpha_c = 1.22\lambda/D_o$ where λ is the *wavelength* of the light. The limit is imposed by diffraction of light by the telescope aperture and is unavoidable. If a telescope is achieving the limit of angular resolution in practice, it is said to be *diffraction-limited*.

line spectra   Spectra which exhibit narrow lines due to absorption (*absorption lines*) or emission (*emission lines*) of *electromagnetic radiation* (called, respectively, *absorption spectra* and *emission spectra*).

liquid   The state of fluid matter characterized by a definite volume but no definite shape.

logarithm (to the base ten)   Given a positive number x, its logarithm to the base ten is the power to which 10 must be raised to obtain the given number. So if $x = 10^y$ then we say that y is the log to the base ten of x, and we write $y = \log_{10}x$.

logarithmic   Pertaining to *logarithms*.

luminosity   The rate at which *energy* in the form of *electromagnetic radiation* leaves a star. It may be quantified by the *absolute magnitude* of the star.

magnitude (astronomical)   A value used to quantify the apparent or absolute brightness of an astronomical object. The magnitude scale is *logarithmic*, such that a difference of 5 magnitudes corresponds to a 100× increase or decrease in brightness. See *apparent magnitude* and *absolute magnitude*.

March equinox   The time at which the centre of the *Sun* is on the *celestial equator*, passing from south to north. On the day of an equinox there are 12 hours between sunrise and sunset all over the Earth. In the Northern Hemisphere this is referred to as the spring equinox or the vernal equinox. It occurs around 21 March.

mass   The quantity of matter in an object. Mass determines the magnitude of the force of gravity acting on a body. This *gravitational force* is also known as the weight of the body. Mass and *energy* are related by Einstein's equation $E = mc^2$.

mass number   The number of *nucleons* (protons and neutrons) in the *nucleus* of an *atom*.

mean   The value obtained by summing all the values in a set of numbers and then dividing the sum by the number of values. In symbols, $\langle x \rangle = \Sigma x_i/n$. See also *median*.

mean solar day   Equal to 24 hours, it is equivalent to the familiar civil day, but about 3 minutes 56 seconds longer than the *sidereal day*. The mean solar day represents the mean length of the *solar days*, averaged over a year.

median   The median value of a set of numbers is the middle one when they are arranged in ascending or descending numerical order. As a measure of the average value, the median is less affected by individually high or low values than is the *mean*.

meridian   An arc that connects the *north celestial pole* and *south celestial pole*, passing through the *zenith*. It is an example of a *great circle*.

meteorites   A type of interplanetary debris that fall to the Earth's surface where they may be collected. They represent material left over from the formation of the Solar System.

microscope   See *petrographic microscope*.

microwaves   The region of the *electromagnetic spectrum* corresponding to *wavelengths* shorter than those of *radio waves* and longer than those of *infrared* radiation.

monochromatic   Used to describe light, or other *electromagnetic radiation*, of a single *wavelength*.

multi-slit spectrograph   A *spectrograph* in which multiple slits are used to obtain the spectra of multiple point-like objects simultaneously. Compare with *fibre-fed spectrograph* and *integral field unit spectrograph*.

neutron (n)   One of the two types of particle in the *nucleus* of an *atom*. Each neutron has an *electric charge* of zero.

newton (N)   The SI unit of force. $1\ \text{N} = 1\ \text{kg m s}^{-2}$.

Newtonian telescope   A *reflecting telescope* incorporating a concave *primary mirror* and a flat, angled *secondary mirror*.

Newton's law of gravitation   States that two particles, of masses m_1 and m_2, separated by a distance r, attract each other with a *gravitational force* that is proportional to the product of their masses, and inversely proportional to the square of their separation. In symbols, the magnitude of the gravitational force on either particle is

$$F = \frac{Gm_1m_2}{r^2}$$

where G, the gravitational constant, is equal to $6.67 \times 10^{-11}\ \text{N m}^2\,\text{kg}^{-2}$.

Newton's second law (of motion)   States that the magnitude of an unbalanced force on an object is equal to the mass of the object multiplied by the magnitude of its acceleration. The direction of the acceleration is the same as the direction of the unbalanced force.

normal distribution   See *Gaussian distribution*.

north celestial pole   The point at which a line from the centre of the Earth through the North Pole intersects the *celestial sphere*.

north point   The point where the north direction in the horizontal plane centred on an observer intersects the *celestial sphere*.

nucleons   The *protons* and *neutrons* in the *nucleus* of an *atom*.

nucleus   The core of an *atom*. It contains nearly all of the mass of the atom and is positively charged. With the exception of the hydrogen nucleus which is a single *proton*, nuclei consist of protons and *neutrons*.

numerical aperture   A measure of a microscope's ability to gather light and resolve fine detail in a specimen. Defined as $NA = n \sin \alpha$ where n is the *refractive index* of the viewing medium, and α is one-half the aperture angle.

objective (*lens* or *mirror*)   The optical element that gathers the light in a telescope. It is also known as the *primary* mirror in a *reflecting telescope*.

objective (*lens*)   The optical element that gathers the light in a microscope, where it usually consists of a multi-element compound lens.

observatory   An establishment with equipment for carrying out astronomical observations. Such a facility may include ground-based telescopes or space-based satellites, and may operate in any part of the *electromagnetic* spectrum. In this book, the term is generally used to refer to a facility that does ground-based optical observations only.

optimal photometry   A technique to optimize the signal-to-noise ratio when carrying out *aperture photometry*. Individual *pixels* are weighted according to their different contributions from the sky and the target object.

overscan strips   Narrow regions of a *CCD* image, usually running down either side, each a few tens of *pixels* wide. They are virtual pixels created by continuing to read out the CCD after the last real pixel has been read out. They indicate how the CCD electronics responds to a zero signal. See *bias frames*.

parsec (pc)   The distance to a celestial body that has a stellar parallax of 1 arcsecond. ($1\ \text{pc} = 3.09 \times 10^{16}\ \text{m}$.)

pass band   The pass band of an astronomical filter is the range of *wavelengths* that are transmitted by it.

path difference   The path difference between two routes is the difference in their lengths. If the two routes join different points in a *coherent* light source to the same point in an image, *constructive interference* or *destructive* *interference* can take place between waves following these routes, depending on whether the path difference is a whole or half number of wavelengths.

petrographic microscope A microscope used for examining thin sections of rocks and minerals. It incorporates two *polarizing filters* that enable the sample to be viewed in *cross-polars*.

phase difference The angle, in terms of 360° times the fraction of a *wavelength*, by which one wave lags or leads another of the same wavelength. If two waves have a phase difference of 0° they are said to be 'in phase', if they have a phase difference of 0.5 cycles or 180° they are said to be 'out of phase'.

photoelectron An *electron* liberated by the absorption of a *photon*.

photometry The technique of measuring the flux of light from astronomical objects.

photon The particle of *electromagnetic radiation* in the photon model of light. *Monochromatic* light consists of photons that each have exactly the same amount of *energy*.

pixel Originally short for 'picture element' but now used to refer both to the individual elements of an image, and to the individual light-sensitive elements within an electronic detector such as a *CCD*.

plane of polarization The plane in which the oscillating electric field of a beam of light, or other *electromagnetic radiation*, is confined.

plane-polarized light *Electromagnetic radiation* in which the oscillating electric field is confined to a single plane, known as the *plane of polarization*. Light from most sources is not polarized – the plane of oscillation of the electric field varies from one part of the beam to another. By passing an un-polarized beam through a *polarizing filter*, a single plane of polarization is selected, and the resulting beam is said to be plane-polarized.

planet A body without its own internal source of luminosity in orbit around a *star*. A planet is much less massive than a star but more massive than an asteroid.

planisphere A device that shows the portion of sky visible from a given location at any time of night, on any night of the year.

plate scale See *image scale*.

point spread function (PSF) The two-dimensional shape of the image produced by a telescope when viewing a distant point source of light. Even in the absence of aberrations and atmospheric turbulence, the PSF will be extended due to *diffraction* by the telescope aperture.

Poisson distribution The asymmetric probability distribution of randomly occurring events as a function of time. As the number of events counted increases, the Poisson distribution becomes more symmetrical and may be approximated by the *Gaussian distribution*.

polar axis In an *equatorial mounting*, this is the axis of the *telescope* that is aligned with the rotation axis of the Earth.

polarized light See *plane-polarized light*.

polarizing filter A sheet of film that allows light with only a single *plane of polarization* to pass through it.

polychromatic Used to describe light, or other *electromagnetic radiation*, composed of a range of *wavelengths*.

population standard deviation (s_n) The square root of the mean of the squares of the deviations of the measured values from their *mean* value. In symbols

$$s_n = \sqrt{\frac{\sum (x_i - \langle x \rangle)^2}{n}}$$

where the mean value $\langle x \rangle$ of the n measurements is $\langle x \rangle = \Sigma x_i/n$. Also known as the root mean square deviation. See *sample standard deviation*.

potential energy *Energy* that is stored, and which depends on the position of an object and not on its motion. So-called because the object has the potential to do work when its position changes. *Gravitational energy* and *electrical energy* are both forms of potential energy.

power The rate at which *energy* is transferred. The SI unit of power is the *watt*.

precession (of the Earth's rotation axis) The Earth's rotation axis is inclined at 23.5° and takes 25 800 years to complete one circuit, so celestial coordinates go through a 25 800 year cycle.

precise Precise measurements are those for which the *random uncertainty* is small.

primary (mirror) The large concave mirror in a *reflecting telescope* which gathers and focuses the light. Also known as the *objective* mirror.

prime focus (of a telescope) The *focal point* of the *primary mirror* of a *reflecting telescope*.

principle of superposition This states that if two or more waves meet in a region of space, then at each instant of time the net disturbance at any point is given by the sum of the disturbances created by each of the waves individually. Gives rise to the phenomenon of *interference*.

prism A (usually) triangular block of glass that may be used to disperse light in a *spectrograph*. Different *wavelengths* of light undergo *refraction* by different angles on entering and emerging from glass, so the colours undergo *dispersion*.

prograde (rotation)   Rotation of an astronomical object such that its spin about its own axis is in the same sense as its orbit around a parent body. The Earth's rotation is prograde since it spins anticlockwise when viewed from above the North Pole and its orbital motion is also anticlockwise when viewed from the same position. See *retrograde*.

proton (p)   One of the two types of particle in the *nucleus* of an *atom*. Each proton has a positive *electric charge*.

radian   An angular measure such that π radians is equivalent to 180°.

radio wave   The region of the *electromagnetic spectrum* corresponding to the very longest *wavelengths* (i.e. lowest energies).

random uncertainty   A random uncertainty will lead to measured values that are scattered in a random fashion over a limited range. Measurements for which the random uncertainty is small are described as *precise*. See *systematic uncertainty* and *Gaussian distribution*.

readout noise   Random noise arising from the CCD electronics as the signal from each pixel is read out. It is typically expressed as a number of electrons per pixel, and may be of order ~10.

real image   An image that may be captured directly on a screen or detector. See also *virtual image*.

reflecting telescope   A telescope that makes use of a concave mirror as its *primary* component.

reflection grating   See *diffraction grating*.

reflective diffraction grating   See *diffraction grating*.

refracting telescope   A telescope that makes use of a convex lens as its primary component.

refraction   The phenomenon whereby the path of a beam of light is altered when the light passes from one medium (e.g. air) to another (e.g. glass) of different *refractive index*. In passing from a medium of *low* refractive index to one of *high* refractive index, light slows down, and a beam is refracted *towards* the normal. Travelling in the other direction, light speeds up, and the beam is refracted *away* from the normal. Because refraction is *wavelength* dependent, *polychromatic* light also undergoes *dispersion* when undergoing refraction. This gives rise to *chromatic aberration* in lenses, and allows *prisms* to be used in *spectrographs*.

refractive index   A measure of the ability of a medium to cause *refraction* of light. It is quantified by the ratio of the *speed of light* in a vacuum to the speed of light in the medium under consideration. Refractive index is therefore always greater than one. The refractive index of a material is usually a function of *wavelength*.

relative photometry   *Aperture photometry* of a target with respect to the flux of light from another object on the same CCD image. The magnitude of the target object is calculated by calibrating against the known *magnitude* of a comparison or reference object. See also *differential photometry*.

relative spectral flux density   The *spectral flux density* at any *wavelength* expressed as a fraction of some reference value. A *continuum normalized spectrum* has a mean relative spectral flux density of one.

resolution (of a microscope objective)   The minimum distance between two points on a sample which can be distinguished as distinct entities. It is related to the *numerical aperture* by $R = 0.61\lambda/NA$ where λ is the *wavelength* of the light. For comparison see *limit of angular resolution* of a telescope.

retrograde (rotation)   Rotation of an astronomical object such that its spin about its own axis is in the opposite sense to its orbit around a parent body. Among the planets of the Solar System, Venus, Uranus and Pluto have retrograde rotations. See *prograde*.

right ascension   The celestial equivalent of longitude, abbreviated as *RA* or α, and measured in hours, minutes, and seconds. Zero hours is at the *First Point of Aries*, and *RA* extends to 24 hours, which is back at the First Point of Aries. At the *celestial equator*, 1 hour of right ascension is equivalent in angular size to 15°.

root mean square (rms) **deviation**   Equivalent to the *population standard deviation*.

sample stage   A rotatable platform used in a microscope to hold the specimen in place. Focusing is achieved by moving the sample stage towards or away from the *objective* lens.

sample standard deviation (s_{n-1})   The square root of: the sum of the squares of the deviations of the measured values from their mean value, divided by one less than the number of values. In symbols

$$s_{n-1} = \sqrt{\frac{\sum (x_i - \langle x \rangle)^2}{n-1}}$$

where the mean value $\langle x \rangle$ of the n measurements is $\langle x \rangle = \Sigma x_i/n$. See *population standard deviation*.

saturate   If the amount of light falling on a given *pixel* of a *CCD* is such that it would give rise to a digital value that is greater than the maximum value that can be stored in the image, the pixel in question is said to be saturated. The analogue-to-digital converters used in CCDs often work at 16-bit precision, so the maximum value that can be stored in any image pixel is $2^{16} - 1 = 65\ 535$.

scattering   The decrease in intensity (of flux density) of radiation as it passes through a medium (e.g. the *interstellar medium* or the Earth's atmosphere) due to the photons being scattered by atoms or molecules such that they emerge in a different part of the spectrum. See *interstellar extinction* and *atmospheric extinction*.

Schmidt–Cassegrain telescope   A *Cassegrain telescope* incorporating a *Schmidt plate*.

Schmidt plate   A correcting lens used in *reflecting telescopes* to correct for the *spherical aberration* inherent in a spherical concave mirror.

Schmidt telescope   A *reflecting telescope* that incorporates a *Schmidt plate*.

scientific notation   A method of writing very large or very small numbers in a concise form. Numbers in scientific notation are expressed as a number between 1 and 10 multiplied by 10 raised to a positive or negative power. For example, the *speed of light* in a vacuum may be written as 2.998×10^8 m s^{-1} and the mass of an electron may be written as 9.109×10^{-31} kg.

secondary (mirror)   The relatively small mirror used in a *Newtonian telescope* or a *Cassegrain telescope* to divert the rays of light from the *primary* mirror towards a detector or *eyepiece*.

secondary extinction coefficient   The secondary extinction coefficient allows for the fact that stars observed at high *airmass* not only appear fainter but also redder. See also *extinction coefficient*.

seeing   Quantifies the quality of the observing conditions at a particular time and site. Numerically, given by the extent of the *point spread function* in an image that has been degraded by the effects of atmospheric turbulence.

September equinox   The time at which the centre of the Sun is on the *celestial equator*, passing from north to south. On the day of an equinox there are 12 hours between sunrise and sunset all over the Earth. In the Northern Hemisphere this is referred to as the autumnal equinox. It occurs around 21 September.

sidereal day   The interval between two successive times at which a given star returns to the same location in the sky. Due to the motion of the Earth around the Sun, the sidereal day is about 3 minutes 56 seconds shorter than the *mean solar day*.

sky flats   *Flat fields* obtained by exposing a *CCD* to the uniformly illuminated twilight sky at dusk or dawn. See *dome flats*.

sky noise   The uncertainty in the number of photons recorded from the sky on a *CCD* image. The sky noise may be combined with the *thermal noise* and accounted for as an overall background noise. If a background aperture on a CCD image contains N_{back} photons per pixel, the background noise is $\sqrt{N_{back}}$ photons per pixel, for a given length of exposure.

slit   Used in conjunction with a *spectrograph*, this is a long, narrow mask placed in the focal plane of a *telescope* where the *polychromatic* (white light) image is first formed. The mask blocks off only the edges of the *image* of a point source, not the top and bottom as well. This enables the light from the image to be dispersed, by a *prism* or *grating*, in a direction perpendicular to that of the slit, such that a *spectrum* is displayed in the image plane of the spectrograph.

solar day   The interval between two successive times at which the Sun returns to the *meridian*. The solar day is longer than the *sidereal day* by a few minutes. Because of the tilt of the Earth's axis and the eccentricity of the Earth's orbit, the solar day is not the same length each day. See also *mean solar day*.

solid   The general form of matter characterized by having a definite volume and shape at fixed *temperature* and pressure.

south celestial pole   The point at which a line from the centre of the Earth through the South Pole intersects the *celestial sphere*.

south point   The point where the south direction in the horizontal plane centred on an observer intersects the *celestial sphere*.

spectral flux density (F_λ)   The rate at which *energy* in the form of radiation is received from a source, per unit area facing the source, per unit *wavelength* range.

spectral order   An integer quantifying how many *wavelengths* of path difference are introduced between successive grooves on a *diffraction grating*. The term appears in the *grating equation*. A diffraction grating will produce diffracted light in several spectral orders, and so give rise to multiple spectra – one corresponding to each spectral order.

spectrograph   A device attached to a telescope used to record the *spectrum* of astronomical objects.

spectroscopy   The study of spectra and spectral lines.

spectrum (plural spectra)   A display of the strength, versus *wavelength* or *frequency*, of the radiation emitted by, or received from a source.

speed of light (in a vacuum)   (*c*) The speed at which *electromagnetic radiation* travels through a vacuum, which, to 3 significant figures, is 3.00×10^8 m s^{-1}.

spherical aberration   An aberration inherent in spherical convex lenses and spherical concave mirrors that causes parallel rays of light passing through different parts of the lens, or reflecting off different parts of the mirror, to be brought to a focus at slightly different points. May be overcome in telescopes using paraboloidal surfaces, or using correcting plates.

spreadsheet   A type of computer program that can be used to manipulate data in rows and columns of cells. The value in a cell can be calculated from a formula, and that formula can involve other cells. Values are recalculated automatically whenever a value in another cell on which it depends changes. Different types of graphs may be plotted to represent the data contained within the various rows and columns.

standard deviation   See *population standard deviation* and *sample standard deviation*.

standard error in the mean   See *uncertainty in the mean*.

standard stars   Well-studied *stars* having a constant and well-defined astronomical *magnitude*.

star   A very large ball of (mainly) hydrogen and helium which is undergoing nuclear fusion reactions in its core. The *Sun* is a typical star.

stop   See *field stop*.

substage light condenser   An adjustable aperture diaphragm and series of lenses in a microscope that is used to gather the light from a *light source* and concentrate it into a cone of light that illuminates the specimen uniformly.

Sun   Our local *star*. It has a mass of 1.99×10^{30} kg and a radius of 6.96×10^8 m.

synchronous rotation   Motion such that a body rotates on its own axis in the same time it takes to orbit another body. The Moon undergoes synchronous rotation with respect to the Earth, and so always presents (almost) the same face towards the Earth.

systematic uncertainty   *Uncertainties* in the measured value of a quantity that cause repeated measurements of that quantity to always differ from the true value in the same way. Measurements in which the systematic

uncertainty is small are described as *accurate*. See *random uncertainty*.

telescope   See *refracting telescope* and *reflecting telescope*.

temperature   A measure of the internal *energy* of a system. The SI unit of temperature is the *kelvin*.

theoretical limit of angular resolution   See *limit of angular resolution*.

thermal noise   An uncertainty introduced into the photon count rate on a *CCD* image as result of the CCD *dark current*. The thermal noise may be combined with the *sky noise* and accounted for as an overall background noise. If a background aperture on a CCD image contains N_{back} photons per pixel, the background noise is $\sqrt{N_{back}}$ photons per pixel.

transformation coefficient   The transformation coefficient adjusts for differences between the equipment used to perform photometric calibration and the equipment used by whoever made the original standard star measurements. It allows for differences in the way the transmissions of filters and the sensitivities of detectors vary with *wavelength*.

transition   The name given to the process by which an *atom* 'jumps' from one state to another with the corresponding emission or absorption of a *photon*.

ultraviolet   The region of the *electromagnetic spectrum* corresponding to *wavelengths* shorter than those of *visible light* and longer than those of *X-rays*.

uncertainties   The possible deviation in a measured value from the true or absolute value. See *systematic uncertainties* and *random uncertainties*.

uncertainty bars   Lines on a graph used to indicate the range of *uncertainties* associated with each particular measurement. Also known as error bars.

uncertainty in the mean   Also known as the standard error in the mean. Given by $\sigma_m = s_n/\sqrt{n}$ or $\sigma_m = s_{n-1}/\sqrt{n-1}$ where s_n is the *population standard deviation* and s_{n-1} is the *sample standard deviation*.

universal time (UT)   Broadly speaking this can be taken to be equivalent to the *civil time* on the Greenwich meridian, without the inclusion of *daylight saving time*. See also *coordinated universal time* (UTC).

virtual image   An image from which light appears to emanate but which may not be captured directly on a screen or detector. The image produced by the *eyepiece* in a telescope or microscope is a virtual image. It is converted into a *real image* by the lens in the eye.

visible light   The region of the *electromagnetic spectrum* corresponding to *wavelengths* shorter than those of *infrared* radiation and longer than those of *ultraviolet* radiation. *Photons* of visible light each have an *energy* of around 2 to 3 *electronvolts*.

volume phase holographic (VPH) diffraction grating   A VPH *diffraction grating* is a transparent medium whose *refractive index* varies from point to point. Closely spaced parallel regions of alternating refractive index can therefore give rise to diffracted light as in a conventional *diffraction grating*. VPH diffraction gratings are increasingly used in *spectrographs* on major telescopes.

watt (W)   The SI unit of *power*. 1 W = 1 J s^{-1}.

wavelength   The distance over which a sinusoidal wave repeats itself, e.g. the distance from one peak of the wave to the next.

wavenumber   A way of characterising *electromagnetic radiation*, defined as one over the *wavelength*. It is most commonly used in *infrared* astronomy, and expressed in cm^{-1}. Consequently, a wavelength of 1 μm corresponds to a wavenumber of 10 000 cm^{-1} and a wavelength of 2 μm corresponds to a wavenumber of 5000 cm^{-1}.

X-ray   The region of the *electromagnetic spectrum* corresponding to *wavelengths* shorter than those of *ultraviolet* radiation and longer than those of *γ-rays*.

zenith   The point at which a vertical line from a given location intersects the *celestial sphere,* in other words the point directly overhead at any instant.

zenith angle   The angle between the *zenith* and the direction to an astronomical object, measured along the *great circle* that passes through the zenith and the point on the *celestial sphere* in the direction to the object. It is equal to 90° minus the *altitude*.

zero-point offset (ζ)   Characterizes the difference between the catalogue magnitude and the instrumental magnitude in the absence of *atmospheric extinction*, as a function of *wavelength*. In order to calibrate photometric observations, the zero-point offset must be determined in each waveband used. The instrumental magnitude (m') and catalogue magnitude (m) of an astronomical object are related by $(m' - m) = \varepsilon X + \zeta$ where ε (the Greek letter epsilon) is the *extinction coefficient* in magnitudes per airmass, X is the *airmass* and ζ (the Greek letter zeta) is the zero-point offset.

INDEX

Glossary terms are in bold. Italics indicate items mainly, or wholly, in a figure or table.